T0205412

Springer Theses

Recognizing Outstanding Ph.D. Research

Aims and Scope

The series "Springer Theses" brings together a selection of the very best Ph.D. theses from around the world and across the physical sciences. Nominated and endorsed by two recognized specialists, each published volume has been selected for its scientific excellence and the high impact of its contents for the pertinent field of research. For greater accessibility to non-specialists, the published versions include an extended introduction, as well as a foreword by the student's supervisor explaining the special relevance of the work for the field. As a whole, the series will provide a valuable resource both for newcomers to the research fields described, and for other scientists seeking detailed background information on special questions. Finally, it provides an accredited documentation of the valuable contributions made by today's younger generation of scientists.

Theses are accepted into the series by invited nomination only and must fulfill all of the following criteria

- They must be written in good English.
- The topic should fall within the confines of Chemistry, Physics, Earth Sciences, Engineering and related interdisciplinary fields such as Materials, Nanoscience, Chemical Engineering, Complex Systems and Biophysics.
- The work reported in the thesis must represent a significant scientific advance.
- If the thesis includes previously published material, permission to reproduce this must be gained from the respective copyright holder.
- They must have been examined and passed during the 12 months prior to nomination.
- Each thesis should include a foreword by the supervisor outlining the significance of its content.
- The theses should have a clearly defined structure including an introduction accessible to scientists not expert in that particular field.

More information about this series at http://www.springer.com/series/8790

Jacob P. Covey

Enhanced Optical and Electric Manipulation of a Quantum Gas of KRb Molecules

Doctoral Thesis accepted by
University of Colorado, Boulder, Colorado, USA

 Springer

Jacob P. Covey
California Institute of Technology
Pasadena, CA, USA

ISSN 2190-5053   ISSN 2190-5061   (electronic)
Springer Theses
ISBN 978-3-030-07452-4   ISBN 978-3-319-98107-9   (eBook)
https://doi.org/10.1007/978-3-319-98107-9

This Springer imprint is published by the registered company Springer Nature Switzerland AG
The registered company address is: Gewerbestrasse 11, 6330 Cham, Switzerland

To Prof. Debbie Jin, my late advisor (jointly with Prof. Jun Ye). We all miss her deeply, and JILA will never quite be the same.

Supervisor's Foreword

Ultracold polar molecules are an idyllic platform for quantum many-body physics and quantum chemistry due to their long-range interactions and rich internal structure. The work described in this thesis constitutes a substantial advance in the capabilities of ultracold polar molecule experiments. Prior to this work, molecular samples were not quantum degenerate, their numbers were limited by chemical reactions, and evidence for many-body physics was lacking. Over the course of this thesis work, molecules were loaded into an optical lattice to control and eliminate collisions and hence chemical reactions. This led to the observation of many-body spin dynamics and the probe of quantum magnetism with spin correlations.

Another substantial advance came from using a quantum synthesis technique based on atomic insulators to increase the filling fraction of the molecules in a three-dimensional optical lattice to 25%. This filling fraction is already above the percolation threshold where long-range spin propagation throughout the 3D lattice is expected. Finally, several limitations to this quantum synthesis approach were identified, and filling fractions towards a molecular insulator are anticipated in future work.

Lastly, this work includes the design, construction, testing, and implementation of a novel apparatus for controlling polar molecules. Soon after Jake Covey joined the KRb experiment, we began to design the second generation apparatus. We were aware of many technical limitations in the first generation JILA system. The design goal of the new apparatus included a substantial increase of atom numbers in both ^{87}Rb BEC and ^{40}K Fermi degenerate gases, precise control of applied electric field and its gradient, and high-resolution imaging for spin-resolved molecular gas microscopy. By the time Jake completed his PhD work in August 2017, the initial characterization of the apparatus had been performed. Only very recently we are now reaping the full benefit of the system: we can now produce over 10^5 ground state ^{40}K^{87}Rb polar molecules, and the road for a deeply degenerate Fermi gas in bulk is clear.

With all of these capabilities successfully demonstrated, we can now combine them to perform evaporative cooling of molecules in two-dimensional geometry to even lower temperatures where p-wave superfluidity may be within reach. There is an enormous growth in ultracold molecule experiments based on ultracold gases in optical lattices, optical tweezer-based molecular assembly, and direct cooling of molecules. All of these efforts will combine optical and electric field control of low-entropy molecular samples. Jake's thesis provides a detailed description of the new system with these capabilities and it could serve as an invaluable reference for people working in this field.

PhD supervisor (jointly with the late Deborah Jin)   Jun Ye
10 June 2018

Acknowledgments

I have had the pleasure of working with many incredible people over the past 6 years. The KRb members I worked with are Amodsen Chotia, Brian Neyenhuis, Bo Yan, Bryce Gadway, Steven Moses, Matt Miecnikowski, Zhengkun Fu, Luigi De Marco, Kyle Matsuda, Will Tobias, and Giacomo Valtolina. I also worked with Kang-Kuen Ni briefly while she was a postdoc working on the JILA eEDM experiment. Although I only overlapped with Amodsen Chotia for a few months, I learned a lot about the KRb experiment from him, and he taught me a lot about experimental physics in general. Brian Neyenhuis was the guiding force on the KRb experiment during my first year, and our initial experiments with molecules in optical lattices were done under his leadership.

The majority of my PhD was done with Steven Moses, Bo Yan, and Bryce Gadway. The four of us were very productive together. Bo is the ultimate experimentalist. He can solve any problem on the experiment, and he always found a way to keep it running well. Bryce has an encyclopedic knowledge of physics, particularly AMO physics. He somehow knows every paper that every group has published. Most of the success during his time on KRb came from his ideas. Steven and I worked together for 5 years. He has an incredible balance of technical skills, physics skills, and problem-solving skills. The latter is incredibly valuable on an experiment that is both aging and extremely complicated. These capabilities combined with his dedication and hardworking mentality have led to a lot of success. I enjoyed working with all three of these guys.

The new KRb team of Luigi De Marco, Kyle Matsuda, Will Tobias, and Giacomo Valtolina is incredibly talented and ambitious. I have full confidence in their capabilities. They all joined the experiment during the second half of the implementation of the new apparatus, and thus they all contributed significantly to its success. Accordingly, they are all well prepared to operate the experiment without me. They breathed new life into the experiment as Steven graduated, Matt and Zhengkun moved on, and Debbie was taken from us. The results that are now coming out of the new apparatus were only possible because of them. I worked with Kyle and Will during their first year as PhD students, and their ability to push the experiment forward while taking three classes has been very impressive. Luigi is a postdoc who did his PhD in physical chemistry studying ultrafast molecular dynamics. He has adapted his skillset to AMO physics remarkably quickly, and his general knowledge of physics and science has been impressive. He brings a bit more depth to the philosophy and direction of the experiment. The newest member is the postdoc Giacomo, who is an AMO expert and has a broad knowledge of ongoing AMO research. The skills and background of Luigi and Giacomo complement each other well.

It has been a pleasure working with Jun Ye, Debbie Jin, and Eric Cornell. The collaborative atmosphere between these three advisors is very constructive and works quite well. The passing of Debbie Jin has been difficult for everyone, but no one was more affected than Jun and Eric. The way they were able to go on afterwards and maintain their groups was incredible. Jun has been an excellent advisor and mentor to me, and he has always been incredibly available and accessible. This in itself is surprising given how busy he is, and he always does his best to help all of his group members and support us in any way we need. Moreover, his optimism and positive attitude has kept us going through the rough patches, and his constant support and belief in us has been very helpful. Debbie Jin was also quite invested in her group members, and she developed very close relationships with many of us. Her friendliness and openness made her loss even harder to bear. Eric's support and advise as a co-advisor had been incredibly useful. He has helped us solve many important problems over the years.

I would be remiss if I did not acknowledge Jun's and Eric's groups for support and advice over the years. Most notably, Jun's Sr experiments are always developing cool new toys and pioneering the best way to do everything. We are indebted to them for help with our Raman cavity, high power distribution and fibers for optical traps, and the new version of our experimental control program. We have benefitted immensely from discussions with Sara Cambell, Ross Hutson, Ed Marti, Shimon Kolkowitz, Aki Goban, Toby Bothwell, and Wei Zhang from Sr; Ming-Guang Hu, Rabin Paudel, Roman Chapurin, and Michael Van De Graaff from the new K apparatus; and Ben Stuhl, Tim Langen, Hau Wu, Dave Reens on the OH Stark decelerator for invaluable advice on AC and DC electric fields.

We have also had a very fruitful collaboration with Ana Maria Rey's group. In particular we worked with Michael Foss-Feig and Kaden Hazzard in the early years, and more recently Michael Wall, Martin Garttner, Arghavan Safavi-Naini, Oscar Acevedo, and Bihui Zhu. They have all done an excellent job being aware of what our experiment is capable of, and giving us space when we need to address technical issues. We attempted the spin-exchange measurements after Kaden and Ana Maria suggested that we would be able to observe them in the lattice even at our relatively low filling fraction at that time.

The JILA instrument shop played a very significant role in much of the work in this thesis, and I have spent a lot of time working with Tracy Keep, Kels Detra, Kim Hagen, Blaine Horner, Todd Asnicar, and Hans Greene. Tracy made the primary contributions to the new apparatus and was very helpful along the entire journey to address every problem that emerged. A lot can change in 6 years, and the JILA instrument shop during my PhD is an excellent example of that. Tracy recently passed away after battling with cancer. Kels left JILA a few years ago. Blaine recently retired. Nevertheless, the JILA instrument shops remain an incredible resource.

The JILA electronics shop and computing team have been invaluable to our experiment. Terry Brown and Carl Sauer are always available to help us with any electrical or electronics problem, particularly high voltage or AC circuits. Their expertise has played a significant role in the electronics of the new apparatus, such as the servos for large electric fields and the rf and microwave coils for K and Rb spin flips. J. R. Raith in computing has been very helpful with all of our computing needs over the years.

On a more personal note, I am thankful to Dan Gresh (Cornell group, eEDM experiment, JILA) for his friendship over the past 6 years and for teaching me the art of powerlifting. I am also grateful to Rory Barton-Grimley (Aerospace Eng., CU-Boulder) for his friendship and for getting me back into hockey. I have enjoyed playing with him and Carrie Weidner, Seth Caliga, and Cam Straatsma (Anderson group, JILA) over the past several years. I would also like to thank my family for their constant support over the past 6 years, especially my wife Jennifer.

Contents

Parts of This Thesis Have Been Published in the Following Journal Articles and Book Chapter

- J.P. Covey, L. De Marco, K. Matsuda, W. Tobias, G. Valtolina, D.S. Jin, J. Ye, A new apparatus for enhanced optical and electric manipulation of ultracold KRb molecules (2018, in preparation)
- J.P. Covey, L. De Marco, O. L. Acevedo, A.M. Rey, J. Ye, An approach to spin-resolved molecular gas microscopy. New J. Phys. **20**, 043031 (2018)
- J.P. Covey, S.A. Moses, D.S. Jin, J. Ye, Controlling ultracold chemical reactions using optical lattices, in *Cold Chemistry: Molecular Scattering and Reactivity Near Absolute Zero* (Royal Society of Chemistry, Cambridge, 2017). ISBN: 978-1-78262-597-1
- S.A. Moses, J.P. Covey, M.T. Miecnikowski, D.S. Jin, J. Ye, New frontiers for quantum gases of polar molecules. Nat. Phys. **13**(1), 13–20 (2017)
- J.P. Covey, S.A. Moses, M. Gärttner, A. Safavi-Naini, M.T. Miecnikowski, Z. Fu, J. Schachenmayer, P.S. Julienne, A. M. Rey, D.S. Jin, J. Ye, Doublon dynamics and polar molecule production in an optical lattice. Nat. Commun. **7**, 11279 (2016)
- S.A. Moses, J.P. Covey, M.T. Miecnikowski, B. Yan, B. Gadway, J. Ye, D.S. Jin, Creation of a low entropy quantum gas of polar molecules in an optical lattice. Science **350**, 6261 (2015)
- K.R.A. Hazzard, B. Gadway, M. Foss-Feig, B. Yan, S.A. Moses, J.P. Covey, N. Yao, M.D. Lukin, J. Ye, D.S. Jin, A.M. Rey, Many-body dynamics of dipolar molecules in an optical lattice. Phys. Rev. Lett. **113**, 195302 (2014)
- B. Zhu, B. Gadway, M. Foss-Feig, J. Schachenmayer, M.L. Wall, K.R.A. Hazzard, B. Yan, S.A. Moses, J.P. Covey, D. S. Jin, J. Ye, M. Holland, A.M. Rey, Suppressing the loss of ultracold polar molecules via the continuous quantum Zeno effect. Phys. Rev. Lett. **112**, 070404 (2014)
- B. Yan, S.A. Moses, B. Gadway, J.P. Covey, K.R.A. Hazzard, A.M. Rey, D.S. Jin, J. Ye, Observation of dipolar spin-exchange interactions with lattice-confined polar molecules. Nature **501**, 521–525 (2013)

- B. Neyenhuis, B. Yan, S.A. Moses, J.P. Covey, A. Chotia, A. Petrov, S. Kotochigova, J. Ye, D.S. Jin, Anisotropic polarizability of ultracold polar $^{40}K^{87}Rb$ molecules. Phys. Rev. Lett. **109**, 230403 (2012)
- A. Chotia, B. Neyenhuis, S.A. Moses, B. Yan, J.P. Covey, M. Foss-Feig, A.M. Rey, D.S. Jin, J. Ye, Long-lived dipolar molecules and Feshbach molecules in a 3D optical lattice. Phys. Rev. Lett. **108**, 080405 (2012)

Chapter 1
Introduction

In this chapter I provide a very brief context of the state-of-the-art in ultracold quantum matter, and how dipolar interactions have become an important tool for many-body physics. I then provide a discussion of the context of my thesis and the topics that it includes.

Polar molecules are an ideal platform for studying quantum information and quantum simulation due to their long-range dipolar interactions. However, they have many degrees of freedom at disparate energy scales and thus are difficult to cool. Ultracold KRb molecules near quantum degeneracy were first produced in 2008. Nevertheless, it was found that even when prepared in the absolute lowest state chemical reactions can make the gas unstable. During my PhD we worked to mitigate these limitations by loading molecules into an optical lattice where the tunneling rates, and thus the chemistry, can be exquisitely controlled. This setting allowed us to start using the rotational degree of freedom as a pseudo-spin, and paved the way for studying models of quantum magnetism, such as the t-J model and the XXZ model. Further, by allowing molecules of two "spin"-states to tunnel in the lattice, we were able to observe a continuous manifestation of the quantum Zeno effect, where increased mobility counterintuitively suppresses dissipation from inelastic collisions. In a deep lattice we observed dipolar spin–exchange interactions, and we were able to elucidate their truly many-body nature. These two sets of experiments informed us that the filling fraction of the molecules in the lattice was only ∼5–10%, and so we implemented a quantum synthesis approach where atomic insulators were used to maximize the number of sites with one K and one Rb, and then these "doublons" were converted to molecules with a filling of 30%. Despite these successes, a number of tools such as high resolution detection and addressing as well as large, stable electric fields were unavailable. Also during my PhD I led efforts to design, build, test, and implement a new apparatus which provides access to these tools and more. We have successfully produced ultracold molecules in this new apparatus, and we are now applying AC and DC electric fields with in vacuum electrodes. This apparatus will allow us to

© Springer Nature Switzerland AG 2018

J. P. Covey, *Enhanced Optical and Electric Manipulation of a Quantum Gas of KRb Molecules*, Springer Theses, https://doi.org/10.1007/978-3-319-98107-9_1

study quantum magnetism in a large electric field, and to detect the dynamics of out-of-equilibrium many-body states.

1.1 Quantum Physics with Ultracold Matter

Since the demonstrations of laser cooling and optical molasses [23, 26, 27] and then evaporative cooling to quantum degeneracy [3, 9], the field of quantum physics with ultracold matter has burgeoned. In the intervening 20 years, a plethora of directions have been pursued, and the capabilities of modern laboratories have been extended beyond anyone's wildest dreams. While the field has grown during this time, most research being pursued within the last decade can be placed into one of the following categories. Researchers are hoping to use ultracold quantum matter to learn more about real-life material systems, such as cuprate high-T_c superconductors; to create new quantum systems that provide enhanced capabilities for measurement; and to create high fidelity quantum bits, or qubits, that can be wired into quantum circuits for quantum information, computation, and communication.

1.1.1 Quantum Simulation

The field of quantum simulation started with bulk, three-dimensional gases of bosonic or fermionic atoms, where research was primarily centered on collective behavior [17, 32] such as vortex formation [1, 21, 25], soliton dynamics [4], and the BCS-BEC (Bardeen-Cooper-Schrieffer to Bose-Einsten Condensate) crossover [11, 15, 28, 35, 36]. From there scientists started reducing the dimensionality using optical lattices [13], which are typically constructed by retroreflecting a laser beam to create a standing wave. Three-dimensional lattices were the natural next step [14, 18], and an enormous amount of quantum simulation work has been done with strongly interacting bosons and fermions in optical lattices [5]. The interactions between the atoms in such experiments are tuned with a Feshbach resonance [7], which will be discussed in detail in Chap. 2.

1.1.2 Quantum Information

While the interactions between neutral atoms can be very strong at short distances, they originate from van der Waals forces and are thus very short range. Conversely, the goal of generating entanglement between many qubits for quantum gate operations requires long-range interactions that can be exquisitely controlled. Such interactions can be derived from electric monopoles, magnetic dipoles, or electric dipoles. Electric monopoles are created using trapped ions, which have been the workhorse of quantum information experiments with cold atoms [22, 33]. Dipolar interactions are the subject of the next section, and they can be generated from highly magnetic atoms [2, 24], polar molecules, or highly excited Rydberg atoms [31].

Trapped ions and Rydberg atoms have been used for quantum gates [22, 31], and the goal of both fields is to scale up the number of qubits for error correction and quantum registers. Both approaches have advantages and disadvantages. For ions, scaling is very difficult because they have such strongly repulsive interactions. Moreover, increasing the dimensionality from the 1D ion chains, which are used almost exclusively, to 2D or 3D crystals while maintaining single-ion addressability and read-out has been prohibitively difficult, although recent attempts look promising [6, 29]. For Rydberg atoms the complexity is not in scaling up the system or changing the dimensionality, but rather entangling larger systems and maintaining high gate fidelities.

1.1.3   Many-Body Quantum Systems out of Equilibrium

Before discussing dipolar interactions in more detail, I would like to describe a new area of research that is becoming incredibly popular in recent years, which is non-equilibrium many-body quantum systems. As the control over neutral atoms continues to improve and as system sizes of strongly interacting particles like ions continue to grow, the two approaches have begun to meet in the middle. Thus, many experiments to date have excellent control of systems of 10–100 particles with strong interactions. Simultaneously, topics like many-body localization [8, 16], quantum thermalization [20], and correlation growth [19, 30] have begun to receive much more attention. Ultracold quantum matter is an excellent template to study such physics because these systems are inherently isolated quantum systems. I mention this because I think these directions are very exciting, and I believe polar molecules will shed a lot more light on these subjects in the future, as I mention later in the thesis.

1.2   Dipolar Interactions

For many research directions, dipolar interactions would be ideal. Firstly, the interactions between electrons in many real materials have a power-law scaling due to screening from the ions forming the lattice, and thus a $1/R^3$ scaling is a reasonable approximation. Secondly, very short-range and very long-range interactions are difficult to work with, as I alluded to above. Therefore, there is a large open area in the interaction strength versus dimensionality parameter space which dipolar particles are particularly suited to explore. Thirdly, dipolar interaction power laws are often the most challenging to simulate theoretically or numerically. Many-body correlations are essential for describing the dynamics, and mean-field treatments are typically not appropriate. From the experimental perspective, however, dipolar systems have been exceedingly difficult to control, as discussed in this work. Nevertheless, important progress has been made.

The difficulty stems from the following facts: (1) highly magnetic atoms have been surprisingly easy to manipulate, but their interactions are relatively weak and not tuneable, (2) Rydberg states are too short-lived and incoherent to realize dynamics, and (3) while polar molecules have strong interactions that are long-lived, they are hard to control. Recent work has been done on Rydberg dressing, which allows Rydberg-state character to be mixed into ground-state atoms, thereby embuing them with long-lived, strong dipolar interactions [12, 34]. I think this is a very promising direction. Nevertheless, this direction is very new and many important questions are so far unanswered. In fact even highly magnetic atoms were largely unexplored when ground-state polar KRb molecules were first made in 2008.

1.3   Topics in This Thesis

In this thesis I will cover all the way from the first creation of ultracold ground-state polar molecules to the present day where we are working towards building a quantum gas microscope of polar molecules in optical lattices for quantum simulation and quantum information. Prior to the work in this thesis the progress of ultracold polar molecules was limited by their complexity. The molecular gases were too hot, the motion in the trap was not entirely controlled, their interaction with the trapping light was poorly understood, and the molecules suffered fast loss from chemical reactions. Further, large electric fields only enhanced the chemical loss in 3D gases. This was only beginning to be controlled by the beginning of my thesis using 2D systems [10]. Moreover, there was a quickly growing number of theoretical proposals for ultracold polar molecule experiments, but many of them required new tools.

In Chap. 2 I will provide an experimental background and overview, where I describe the setting in which polar molecules were first created. Then I will describe the physics of so-called Feshbach molecules and how a Raman laser sequence can be used to convert Feshbach molecules to the rovibronic ground state where they have large dipole moments. Then I will provide an overview of the apparatus, specifically the first generation apparatus and the laser systems that we used in both the old and new experiments.

In Chap. 3 I will outline the physics and chemistry of ultracold molecules that was observed in bulk gases in 2D and 3D systems. While I was not directly involved in these experiments, I will describe them because they are important to set the stage for the work I was involved in with optical lattices, as well as future work on evaporative cooling and stabilized dipoles in large electric fields. Moreover, I coauthored a chapter for a book entitled "Low energy and low temperature molecular scattering" in which we review all of these topics.

In Chap. 4 I will introduce molecules in optical lattices, and the stability they afford. I will give an overview of the physics of optical lattices, band structure, superfluids, and Mott insulators. Then, I will describe all the experiments we have performed to understand molecules in optical lattices and the effects of optical traps which are unique to polar molecules.

In Chap. 5 I will present our experiments studying dipolar spin–exchange interactions between molecules in two rotational states pinned in a deep optical lattice. These observations include many-body dynamics, and constitute an important first step towards quantum magnetism with polar molecules.

In Chap. 6 I will describe our extended efforts to reduce the entropy, or temperature, of our molecular sample. Ultimately, our success came using a quantum synthesis approach of atomic insulators in the lattice that we convert to molecules with high fidelity. This approach has allowed us to make the first low-entropy quantum gas of polar molecules, which could be considered a quantum degenerate sample in a 3D optical lattice, and could allow for novel experiments on many-body out-of-equilibrium spin systems.

In Chap. 7 I will describe the new generation KRb experiment and all the design features that will present enormous improvements to the old apparatus. Specifically, it is designed for large, tunable electric fields generated by in-vacuum electrodes, as well as high resolution detection and addressing. Such tools are necessary to take the next steps with capabilities described in the prior two chapters: quantum magnetism and spin physics, and a low-entropy system where exotic, many-body dynamics and phases emerge.

In Chap. 8 I will report how the new apparatus was designed, built, and tested. There were an enormous number of technical limitations that must be overcome, and a huge number of tests were required to successfully build the new apparatus machine to provide all the tools described in the previous chapter. This chapter is entirely technical, but it is by far the longest chapter in this thesis. I hope this fact illustrates just how complex the new apparatus is, and how exciting our ultimate success has been upon the completion of its implementation.

In Chap. 9 I will outline the procedure for producing ultracold atoms and molecules in the new apparatus, and describe the conditions that we have achieved. These conditions are quite comparable to the old chamber, despite a significantly more complicated experimental procedure.

In Chap. 10 I will present the novel experiments that we are performing in the new apparatus, taking advantage of all the new tools for which it was designed. We were immediately able to go beyond anything that we were capable of in the old apparatus, such as large electric fields, polarization-selectivity of rotational transitions, and high resolution imaging.

In Chap. 11 I will close with an outlook of where we have come and what the next steps are for polar molecules, and the new apparatus in particular. The new apparatus opens the door to many experiments which have been proposed over the years, and I will highlight the ideas that are next on our list.

References

1. J.R. Abo-Shaeer, C. Raman, J.M. Vogels, W. Ketterle, Observation of vortex lattices in bose-einstein condensates. Science **292**(5516), 476–479 (2001)
2. K. Aikawa, A. Frisch, M. Mark, S. Baier, R. Grimm, F. Ferlaino, Reaching fermi degeneracy via universal dipolar scattering. Phys. Rev. Lett. **112**, 010404 (2014)
3. M.H. Anderson, J.R. Ensher, M.R. Matthews, C.E. Wieman, E.A. Cornell, Observation of Bose-Einstein condensation in a dilute atomic vapor. Science **269**(5221), 198–201 (1995)
4. B.P. Anderson, P.C. Haljan, C.A. Regal, D.L. Feder, L.A. Collins, C.W. Clark, E.A. Cornell, Watching dark solitons decay into vortex rings in a bose-einstein condensate. Phys. Rev. Lett. **86**, 2926–2929 (2001)
5. I. Bloch, J. Dalibard, S. Nascimbene, Quantum simulation with ultracold quantum gases. Nat. Phys. **8**, 267–276 (2012)
6. J.W. Britton, B.C. Sawyer, A.C. Keith, C.-C. Joseph Wang, J.K. Freericks, H. Uys, M.J. Biercuk, J.J. Bollinger, Engineered two-dimensional Ising interactions in a trapped-ion quantum simulator with hundreds of spins. Nature **484**, 489–492 (2012)
7. C. Chin, R. Grimm, P. Julienne, E. Tiesinga, Feshbach resonances in ultracold gases. Rev. Mod. Phys. **82**, 1225–1286 (2010)
8. J.-Y. Choi, S. Hild, J. Zeiher, P. Schauß, A. Rubio-Abadal, T. Yefsah, V. Khemani, D.A. Huse, I. Bloch, C. Gross, Exploring the many-body localization transition in two dimensions. Science **352**(6293), 1547–1552 (2016)
9. K.B. Davis, M.O. Mewes, M.R. Andrews, N.J. van Druten, D.S. Durfee, D.M. Kurn, W. Ketterle, Bose-Einstein condensation in a gas of sodium atoms. Phys. Rev. Lett. **75**, 3969–3973 (1995)
10. M.H.G. de Miranda, A. Chotia, B. Neyenhuis, D. Wang, G. Quéméner, S. Ospelkaus, J.L. Bohn, J. Ye, D.S. Jin, Controlling the quantum stereodynamics of ultracold bimolecular reactions. Nat. Phys. **7**(6), 502–507 (2011)
11. B. DeMarco, D.S. Jin, Onset of fermi degeneracy in a trapped atomic gas. Science **285**(5434), 1703–1706 (1999)
12. C. Gaul, B.J. DeSalvo, J.A. Aman, F.B. Dunning, T.C. Killian, T. Pohl, Resonant rydberg dressing of alkaline-earth atoms via electromagnetically induced transparency. Phys. Rev. Lett. **116**, 243001 (2016)
13. A. Görlitz, J.M. Vogels, A.E. Leanhardt, C. Raman, T.L. Gustavson, J.R. Abo-Shaeer, A.P. Chikkatur, S. Gupta, S. Inouye, T. Rosenband, W. Ketterle, Realization of bose-einstein condensates in lower dimensions. Phys. Rev. Lett. **87**, 130402 (2001)
14. M. Greiner, O. Mandel, T. Esslinger, T.W. Hansch, I. Bloch, Quantum phase transition from a superfluid to a Mott insulator in a gas of ultracold atoms. Nature **415**(6867), 39–44 (2002)
15. M. Greiner, C.A. Regal, D.S. Jin, Emergence of a molecular Bose-Einstein condensate from a Fermi gas. Nature **426**, 537–540 (2003)
16. D.A. Huse, R. Nandkishore, V. Oganesyan, Phenomenology of fully many-body-localized systems. Phys. Rev. B **90**, 174202 (2014)
17. D.S. Jin, J.R. Ensher, M.R. Matthews, C.E. Wieman, E.A. Cornell, Collective excitations of a bose-einstein condensate in a dilute gas. Phys. Rev. Lett. **77**, 420–423 (1996)
18. R. Jördens, N. Strohmaier, K. Günter, H. Moritz, T. Esslinger, A mott insulator of fermionic atoms in an optical lattice. Nature **455**, 204–207 (2008)
19. P. Jurcevic, B.P. Lanyon, P. Hauke, C. Hempel, P. Zoller, R. Blatt, C.F. Roos, Quasiparticle engineering and entanglement propagation in a quantum many-body system. Nature **511**, 202–205 (2014)
20. A.M. Kaufman, M.E. Tai, A. Lukin, M. Rispoli, R. Schittko, P.M. Preiss, M. Greiner, Quantum thermalization through entanglement in an isolated many-body system. Science **353**(6301), 794–800 (2016)
21. A.E. Leanhardt, A. Görlitz, A.P. Chikkatur, D. Kielpinski, Y. Shin, D.E. Pritchard, W. Ketterle, Imprinting vortices in a bose-einstein condensate using topological phases. Phys. Rev. Lett. **89**, 190403 (2002)

22. D. Leibfried, R. Blatt, C. Monroe, D. Wineland, Quantum dynamics of single trapped ions. Rev. Mod. Phys. **75**, 281–324 (2003)
23. P.D. Lett, R.N. Watts, C.I. Westbrook, W.D. Phillips, P.L. Gould, H.J. Metcalf, Observation of atoms laser cooled below the doppler limit. Phys. Rev. Lett. **61**, 169–172 (1988)
24. M. Lu, N.Q. Burdick, S.H. Youn, B.L. Lev, Strongly dipolar bose-einstein condensate of dysprosium. Phys. Rev. Lett. **107**, 190401 (2011)
25. M.R. Matthews, B.P. Anderson, P.C. Haljan, D.S. Hall, C.E. Wieman, E.A. Cornell, Vortices in a bose-einstein condensate. Phys. Rev. Lett. **83**, 2498–2501 (1999)
26. G. Modugno, C. Benkő, P. Hannaford, G. Roati, M. Inguscio, Sub-doppler laser cooling of fermionic ^{40}K atoms. Phys. Rev. A **60**, R3373–R3376 (1999)
27. W. Petrich, M.H. Anderson, J.R. Ensher, E.A. Cornell, Behavior of atoms in a compressed magneto-optical trap. J. Opt. Soc. Am. B **11**(8), 1332–1335 (1994)
28. C.A. Regal, M. Greiner, D.S. Jin, Observation of resonance condensation of fermionic atom pairs. Phys. Rev. Lett. **92**, 040403 (2004)
29. P. Richerme, Two-dimensional ion crystals in radio-frequency traps for quantum simulation. Phys. Rev. A **94**, 032320 (2016)
30. P. Richerme, Z.-X. Gong, A. Lee, C. Senko, J. Smith, M. Foss-Feig, S. Michalakis, A.V. Gorshkov, C. Monroe, Non-local propagations of correlations in quantum systems with long-range interactions. Nature **511**, 198–201 (2014)
31. M. Saffman, T.G. Walker, K. Mølmer, Quantum information with rydberg atoms. Rev. Mod. Phys. **82**, 2313–2363 (2010)
32. D.M. Stamper-Kurn, H.-J. Miesner, S. Inouye, M.R. Andrews, W. Ketterle, Collisionless and hydrodynamic excitations of a bose-einstein condensate. Phys. Rev. Lett. **81**, 500–503 (1998)
33. Q.A. Turchette, C.S. Wood, B.E. King, C.J. Myatt, D. Leibfried, W.M. Itano, C. Monroe, D.J. Wineland, Deterministic entanglement of two trapped ions. Phys. Rev. Lett. **81**, 3631–3634 (1998)
34. J. Zeiher, R. van Bijnen, P. Schauß, S. Hild, J.-Y. Choi, T. Pohl, I. Bloch, C. Gross, Many-body interferometry of a rydberg-dressed spin lattice. Nat. Phys. **12**, 1095–1099 (2016)
35. M.W. Zwierlein, C.A. Stan, C.H. Schunck, S.M.F. Raupach, S. Gupta, Z. Hadzibabic, W. Ketterle, Observation of Bose-Einstein condensation of molecules. Phys. Rev. Lett. **91**, 250401 (2003)
36. M.W. Zwierlein, J.R. Abo-Shaeer, A. Schirotzek, C.H. Schunck, W. Ketterle, Vortices and superfluidity in a strongly interacting fermi gas. Nature **435**, 1047–1051 (2005)

Chapter 2
Experimental Background and Overview

In this chapter I present an introduction to the field of ultracold polar molecules, which is less than a decade old. By extending the techniques that have been developed for ultracold atoms, I describe how a molecular quantum gas can be produced. I then describe the first generation JILA KRb apparatus, it's move into the new X-wing of JILA in 2012, and the laser systems that have been in use on both the first and second generation KRb machines since the move.

2.1 Creation of Ultracold Molecules

The field of ultracold polar molecules has exploded over the past decade as researchers have worked to extend the control offered in ultracold atom experiments producing atomic Bose-Einstein condensates (BECs) [2, 5, 6, 16] and Degenerate Fermi gases (DFGs) [19, 20, 53] to the world of molecules. Heteronuclear molecules can have intrinsic electric dipole moments, and they provide strong, long-range, and anisotropic interactions with precise tunability. A strong interest has emerged in the scientific community to study systems with long-range interactions, as discussed in the previous chapter. These systems are ideal candidates for the study of strongly correlated quantum phenomena, as well as for quantum simulation of lattice models relevant to some outstanding problems in condensed-matter physics. The anisotropic nature of dipolar interactions provides powerful opportunities to control chemical reactions in the low energy regime, and dipolar interactions can also give rise to novel forms of quantum matter such as Wigner crystallization [31], d-wave superfluidity in optical lattices [33], fractional Chern insulators [58], and spin-orbit coupling [51].

Efforts to produce stable gases of ultracold polar molecules started in the early 2000s, and are documented in a recent review [8]. However, the production of ground-state polar molecules in the quantum regime, where motional degrees of

© Springer Nature Switzerland AG 2018
J. P. Covey, *Enhanced Optical and Electric Manipulation of a Quantum Gas of KRb Molecules*, Springer Theses, https://doi.org/10.1007/978-3-319-98107-9_2

Fig. 2.1 Potential energy
wells of the closed and open
channel near a Feshbach
resonance. The energy offset
E_c can be adjusted via a
differential Zeeman shift
using a magnetic field to
realize resonant coupling of
the two channels. Reproduced
from Ref. [9]

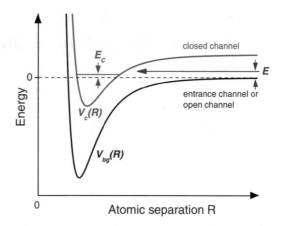

freedom must be described quantum mechanically, was a challenging task and it took a major effort to achieve the first success in 2008 [41]. The key was the combination of the use of a Feshbach resonance for magneto-association of bialkali atoms and coherent optical state transfer via stimulated Raman adiabatic passage (STIRAP) [35, 54].

Feshbach resonances, where the energy of a colliding pair of free atoms matches the energy of a bound state associated with a molecular potential (see Fig. 2.1), have been studied extensively in the context of controlling atomic interactions via magnetic fields, and they have become one of the most powerful tools for quantum gas experiments. Homonuclear, weakly-bound molecules associated from free atoms using a Feshbach resonance (i.e. Feshbach molecules) were produced directly from single-species ultracold atomic samples [21, 27, 29]. In particular, homonuclear Feshbach molecules, even if composed of fermionic atoms, are themselves bosons, and their production enables studies of a crossover from the BCS-type Cooper pairing of fermions into a BEC of weakly-bound molecules [3, 25, 61]. In an optical lattice, such weakly-bound molecules can be effectively protected: when the tunneling rate is smaller than the on-site reaction rate, molecular tunneling is suppressed via the quantum Zeno effect [50]. Work on using the STIRAP technique to transfer molecules to more deeply bound states was performed on both Rb_2 [34, 56] and Cs_2 molecules [13, 14]. However, homonuclear molecules do not have an intrinsic electric dipole moment, and are thus not useful for studying dipolar scattering or long-range dipole–dipole interactions.

Inspired by the work with weakly bound homonuclear molecules, many researchers started producing heteronuclear dimers, involving various combinations of alkali atoms [9, 32]. However, the path to deeply bound, or ground-state, molecules was challenging. The earliest efforts for producing cold polar molecules relied on photoassociation to optically couple free atoms to an excited molecular electronic state, which then spontaneously decayed to many rovibrational states in the ground electronic potential. In order to collect the population in the rovibrational ground state, one can choose an excited state with good Franck-

Condon overlap [11, 23] with both the free atom state and the ground rovibrational state [30]. However, the efficiencies for both the excitation from free atoms to an excited molecular state and the subsequent spontaneous decay to a specific rovibrational state in the ground are very low. As a consequence, the phase-space density (particle density in coordinate and momentum space) of such cold molecule samples is typically below $\sim 10^{-10}$ [28]. Several more recent experiments have managed to improve on this limitation by using electronically excited states with stronger Franck-Condon coupling to deeply bound states (see, e.g., References [47, 49]), but they are not in the rovibrational ground state, and the phase-space densities are still very far from reaching the quantum regime. For reference, the phase-space-densities routinely reached with atomic quantum gases are ≥ 1, and quantized motions, quantum statistics, and collective behavior become apparent.

Alternatively, the coherent transfer approach first adiabatically converts a pair of free atoms of different species into a highly vibrationally excited state in the ground electronic potential using a Feshbach resonance, and then couples these Feshbach molecules to the rovibrational ground state via an adiabatic Raman transfer process through an intermediate electronic excited state. This approach effectively maintains coherence from the initial to final quantum state, and thus introduces a minimal amount of entropy during the atom-to-molecule conversion process. The first success came in 2008 with the production of heteronuclear fermionic KRb molecules [41, 45]. Research on deeply bound heteronuclear molecules has exploded in recent years. Several groups have now produced heteronuclear molecules in their deeply bound ground state, in species such as bosonic RbCs [36, 52], fermionic NaK [46], and bosonic NaRb [26]. The fermionic KRb molecules recently reached low entropies in an optical lattice [12, 38], to be described in Chap. 6.

2.1.1 Magneto-Association

A pair of free atoms can be directly converted to a weakly bound molecule in the ground potential when this bound molecular state (associated with a closed channel) becomes degenerate in energy with the continuum state of two free atoms (associated with an open channel). This type of Feshbach resonance exists for essentially all atomic species with a magnetic moment, and can be used to tune the relative energy between the atomic and molecular state. The two-channel model for a Feshbach resonance is shown in Fig. 2.1. Since the two channels typically have different magnetic moments, the energy difference between them, E_c, can be adjusted with a magnetic field to realize resonant coupling of the two channels [9]. As a result of this resonant coupling, the scattering length acquires a dependence on magnetic field, shown in Fig. 2.2a, and given by

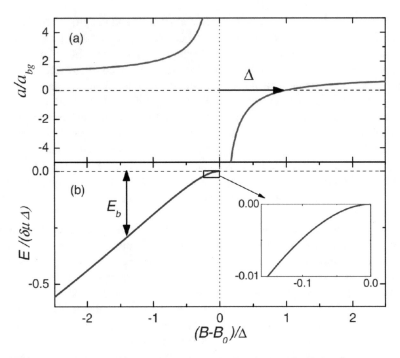

Fig. 2.2 Scattering length and energy of two atoms in the vicinity of a Feshbach resonance. (**a**) shows the scattering length normalized to the background scattering as a function of field, while (**b**) shows the binding energy as a function of field. Note that this figure assumes $a_{bg} > 0$, which is not always the case. Reproduced from Ref. [9]

$$a(B) = a_{bg}\left(1 - \frac{\Delta}{B - B_0}\right), \tag{2.1}$$

where a_{bg} is the background scattering length that parameterizes the interaction strength, and Δ is the magnetic field width of the resonance, which is the difference between B_0 and the field that corresponds to $a = 0$ (see Fig. 2.2a). Note that the energy scale plotted in Fig. 2.2b is normalized by $\delta\mu$, which is the difference in magnetic moments of the two channels. Typical widths in the most experimentally used resonances are $\sim$1–300 G, although some molecules (e.g., Cs_2 [14, 27]) use a resonance width of several mG.

At the field B_0, the energy difference between the closed and open channels vanishes, and the resultant scattering length of the two atoms diverges. Tuning B away from B_0 on the side of the positive scattering length leads to a molecular binding energy E_b, and it is given by

$$E_b = \hbar^2/2m_R a^2, \tag{2.2}$$

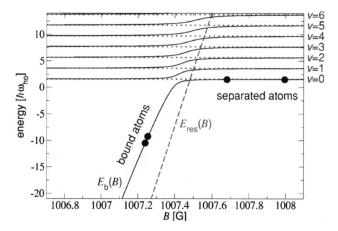

Fig. 2.3 The energy of two atoms as a function of magnetic field. The harmonic oscillator levels of the atoms in a harmonic trapping potential are shown above the resonance. The closed channel energy is $E_{\mathrm{res}}(B)$. The lowest harmonic oscillator level couples to the Feshbach molecule, whose energy reduces below the resonance as $E_{\mathrm{b}}(B)$. Reproduced from Ref. [32]

where m_{R} is the reduced mass of the atom pair and a is the scattering length of the atoms [9]. Typical binding energies in Feshbach molecule experiments are ~ 0.1– 1 MHz.

One way to produce Feshbach molecules is to sweep B across the resonance from the attractive side ($a_{\mathrm{bg}} < 0$) to the repulsive side ($a_{\mathrm{bg}} > 0$) so that the pair of free atoms adiabatically enters the closed molecular channel. Figure 2.3 illustrates this process schematically, and also shows the change of the molecular binding energy as a function of B. The free atoms are confined in a harmonic trap, and the different entrance levels correspond to different harmonic oscillator states in the trap. In the limit that the sweep is adiabatic and the harmonic oscillator states are well resolved, only the lowest energy free atomic state couples to the bound channel. This process is well described by a Landau-Zener coupling mechanism, and the probability that free atoms are adiabatically converted to molecules is given by [32]:

$$P_{\mathrm{FbM}} = 1 - e^{-2\pi \delta_{\mathrm{LZ}}}. \tag{2.3}$$

Here δ_{LZ} is the Landau-Zener parameter, which in a weak harmonic trap is [32]

$$\delta_{\mathrm{LZ}}^{\mathrm{ho}} = \frac{\sqrt{6}\hbar}{\pi\, 2\phi l_{\mathrm{ho}}^3} \left| \frac{a_{\mathrm{bg}}\Delta}{\dot{B}} \right|, \tag{2.4}$$

where $l_{\mathrm{ho}} = \sqrt{\hbar/m\bar{\omega}}$ is the harmonic oscillator length of the molecule in the harmonic trap (with mean trap frequency $\bar{\omega}$), and $\dot{B} = dB/dt$ is the sweep rate across the resonance.

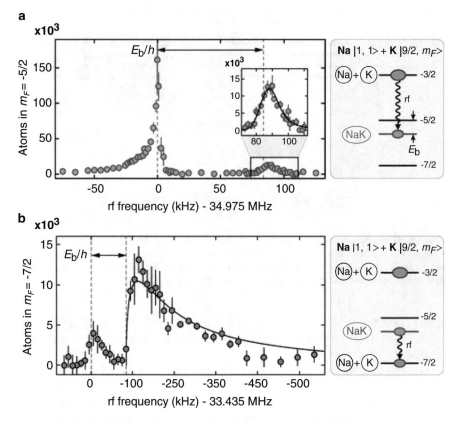

Fig. 2.4 (**a**) RF association, and (**b**) RF dissociation of NaK Feshback molecules, using two different hyperfine states of K. In both cases the binding energy can be seen as a separate peak that can be spectroscopically resolved from the free K atom peak. Reproduced from Ref. [57]

An alternative method to create Feshbach molecules is to couple the open and closed channels with a radio frequency (rf) field. This technique is used in many experiments, and enables coherent Rabi oscillations between free atoms and molecules [21, 42]. Figure 2.4 shows a beautiful illustration of this technique.

2.1.2 Coherent Optical Transfer

After producing weakly bound Feshbach molecules, a two-photon coherent population transfer technique, STImulated Raman Adiabatic Passage (STIRAP) [35, 54], is used to transfer molecules to the rovibrational ground state. Three states are involved in the process: the initial Feshbach molecule state $|i\rangle$, an electronically excited state $|e\rangle$ (which is usually short-lived), and the rovibrational ground state

$|g\rangle$. The excited state is chosen, after an exhaustive spectroscopic search, to optimize the Franck-Condon factors with both $|i\rangle$ and $|g\rangle$. The states are coupled with two laser fields: the up leg, with Rabi frequency Ω_1, couples $|i\rangle$ and $|e\rangle$; the down leg, with Rabi frequency Ω_2, couples $|g\rangle$ and $|e\rangle$. The transition dipole moments (including the Franck-Condon factor) for both up and down legs are 0.5 to 1×10^{-2} atomic units [41]. The goal of STIRAP is to maintain a field-dressed dark state $\cos\theta|i\rangle + \sin\theta|g\rangle$, where $\theta = \tan^{-1}\left(\frac{\Omega_1}{\Omega_2}\right)$. The laser intensities should adiabatically change from $\Omega_2/\Omega_1 \gg 1$ to $\Omega_1/\Omega_2 \gg 1$. In practice, the laser intensities are ramped over a duration $\tau \gg 1/\Omega_{1,2}$. Since the population is trapped in the dark state, population transfer takes place from $|i\rangle$ to $|g\rangle$ without populating the short-lived state $|e\rangle$. The typical STIRAP protocol for ultracold molecules is based on a dark resonance configuration, where both the one-photon and the two-photon detunings are zero. The excited state population is therefore adiabatically eliminated under these conditions, provided $\beta \ll 1/\tau$, where β is the relative laser linewidth of the two-photon transition [59]. Thus, the two laser fields must be phase coherent with each other during the time τ. For typical pulse times $\sim 10\,\mu s$, relative laser linewidths less than $1\,kHz$ are required, which can be achieved by stabilizing the lasers to an optical frequency comb [41, 45] or a high-finesse optical cavity [1, 24].

Once in the ground state, we can remove the unpaired atoms by sending in resonant light on the K and Rb cycling transitions. Then, by reversing the order of the intensity ramps, the ground-state molecules are converted via STIRAP back to Feshbach molecules. The Feshbach molecules are then dissociated and the resulting atoms are imaged on either the K or Rb cycling transition. It is also possible to directly image the ground-state molecules, but the lack of optical cycling transitions limits the signal-to-noise ratio [55].

Figure 2.5 shows the states used for efficient production of ground-state KRb molecules. Bi-alkali molecules have orbital angular momentum $\Lambda = 0$ in their ground molecular potentials (e.g., Σ states), and are thus described by Hund's Case (b) [7]. The initial Feshbach state is the most weakly bound vibrational level of the $^3\Sigma^+$ ground potential, while the rovibrational ground state is in the $^1\Sigma^+$ ground potential. Thus, the state $|e\rangle$ must contain an admixture of singlet and triplet characters in order to allow reasonable transition strengths on both legs of the transfer. Such an admixture results from significant spin-orbital coupling in the excited electronic state. Using this scheme, we typically find a $\sim 90\%$ one-way transfer efficiency from the initial Feshbach molecule state to the ground state (since the coherent process is fully reversible, we have the same efficiency from the ground state back to the Feshbach state). An important aspect to consider is that the difference in energy between the initial and final states is $\sim 4000\,K$, which is enormous compared to the temperature of the gas ($\sim 100\,nK$). Therefore, one might wonder whether any of this extra energy is deposited into the kinetic energy of molecules, which would heat the molecular gas out of the ultracold regime. However, since the transfer process is fully coherent, the energy difference between the states is carried away by the photons, and the gas remains in the $\sim 100\,nK$ regime. However, because the Raman lasers have different wavelengths,

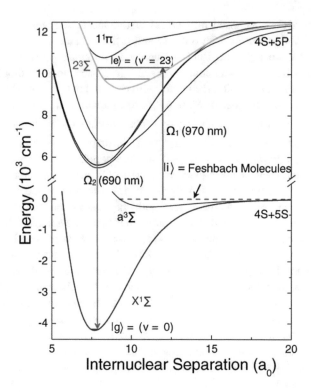

Fig. 2.5 Two-photon coherent state transfer for KRb from the weakly bound Feshbach molecule state $|i\rangle$ to the absolute molecular ground state $|f\rangle$ ($v = 0$ (vibration), $N = 0$ (rotation) of $X^1\Sigma$). Reproduced from Ref. [41]

the molecules do get a momentum kick equal to the difference of the photon momenta [43]. When producing molecules directly in an optical lattice (to be discussed later), this process allows the molecules to occupy the ground band of the lattice with very high probability in typical conditions of our experiment, e.g. lattice depth, etc. [15, 38, 48].

2.2   The First Generation KRb Machine

In this section I will describe the basics of the first generation KRb machine. More details can be found primarily in the thesis by Josh Zirbel [60], but also Kang-Kuen Ni [40], Marcio de Miranda [17], Brian Neyenhuis [39], and Steven Moses [37]. Therefore I will only describe what I think is important to contrast with the second generation apparatus which I will be describing in later chapters.

The first generation apparatus is based on a double chamber design with a long differential pumping tube between, as shown in Fig. 2.6. The first generation

Fig. 2.6 The old vacuum chamber in the first generation KRb experiment. The vapor cell MOT area is on the right, and separated by a large differential pumping section and a gate valve is the science cell side

chamber has K dispersers, which are enriched to 5% ^{40}K (natural abundance is 0.012%), and Rb ampules which we cool with a thermo-electric controller (TEC). In this chamber we load a dual MOT (Magneto-Optical Trap) of K and Rb from a background vapor of both species. The pressure is maintained by controlling the temperatures of the K dispenser and Rb ampule, and we typically operated with a pressure $\sim 10^{-10}$ mbar, which corresponded to a MOT fill time and lifetime of ~ 5 s. The MOT sizes were typically 2–3×10^9 for Rb and 1–2×10^7 for K. Upon loading the MOT, we performed a compressed MOT stage for both K and Rb. Then we did bright optical molasses (we now use grey optical molasses for both species), and finally optical pumping to the stretched, magnetically trappable states.

After loading the quadrupole trap we transfer atoms through the differential pumping tube to the science cell by translating the anti-Helmholtz coils on a motion track. The transfer efficiency is $\sim 50\%$ for Rb and $\sim 20\%$ for K. The temperature after transfer is roughly the same for both, but the temperature of K is much higher than Rb before transfer. The differential pumping tube actually shaves down the K cloud and thus reduces the temperature. The tube actually contains a strange "bump" region in the differential pumping tube that is the primary limitation to the transfer efficiency (its original intention was to prevent the migration of alkali metal from the hotter MOT region to the room temperature science region). This bump region is described in Josh Zirbel's and Kang-Kuen Ni's thesis, so I will just say that it is an unnecessary complication which reduces the transfer efficiency. The need for a new apparatus, to be described later, was initially recognized before my arrival partially with the goal of removing this bump.

Once we reach the science cell, the atoms are loaded into a Ioffe-Pritchard (IP) trap (shown in Fig. 2.7) where Rb undergoes forced rf-knife evaporative cooling by driving the $|F, m_F\rangle = |2, 2\rangle$ to $|1, 1\rangle$ transition (where F is the total atomic angular momentum and m_F is its projection onto the magnetic field axis). K is sympathetically cooled by Rb during the evaporation. Note that we continually remove Rb atoms that may appear in the $|2, 1\rangle$ state, which will undergo inelastic collisions with K. This is done by using a so-called "$|2, 1\rangle$-cleaner"

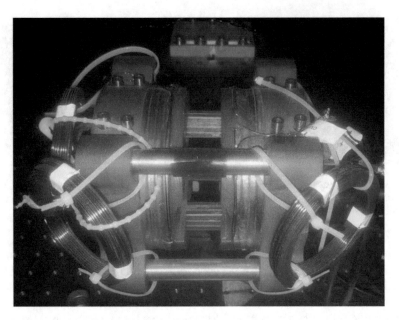

Fig. 2.7 The phenolic form that holds the Ioffe-Pritchard trap around the science cell of the old chamber

which continuously drives the $|2, 1\rangle$ to $|1, 0\rangle$ transition resonant with the bottom of the trap during the evaporation. This evaporation starts with 2–3 $\times$ 10^8 Rb and 2×10^6 K atoms at a temperature of $\sim$300–400 μK and ends with 4×10^6 Rb and 6–7 $\times$ 10^5 K atoms at a temperature of $\sim$1 μK. The trap frequencies in the IP trap are $\omega_z = 2\pi \cdot 20$ Hz and $\omega_r = 2\pi \cdot 156$ Hz for Rb in $|2, 2\rangle$.

From these conditions we loaded atoms into a crossed optical dipole trap of beams with $1/e^2$ radii of 40×200 μK. Typical powers before optical evaporation were $\sim$3 W in each beam, and then they lower over 2–3 s to evaporate Rb against gravity, again sympathetically cooling K. This would yield typical conditions of 3×10^5 Rb and 3×10^5 K at 300 nK, just above T_c and T_F, the Bose-Einstein transition temperature and the Fermi temperature of Rb and K, respectively. Such conditions yield the highest number of Feshbach molecules and ground state molecules, and we were routinely able to get 3–4 $\times$ 10^4 at $T = T_F$. However, for the later discussion of a low-entropy quantum gas in an optical lattice we made deeply degenerate Bose-Fermi mixtures, and we could get 2–3 $\times 10^5$ Rb at 0.1 T_c and 2–3 $\times$ 10^5 K at 0.3 T_F. This improvement in the phase-space-density was obtained both by evaporating further in the optical trap and by increasing the aspect ratio of the optical traps.

The glass cell on the old chamber was an uncoated pyrex cuvette that was fused to a glass-to-metal seal. The assembly and fusing process was done at JILA by Hans Greene, but resulted in very large surface features such as ripples in the glass. Therefore, it was never possible to generate precise optical potentials or to perform

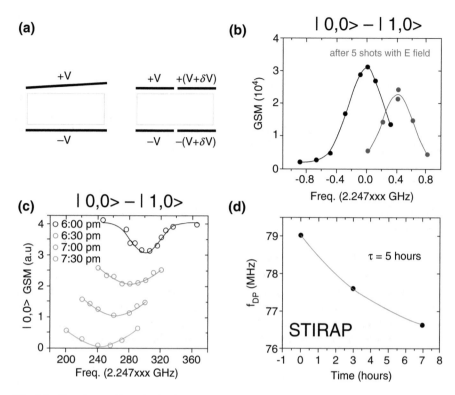

Fig. 2.8 Charging problems in the first generation apparatus. (**a**) shows the glass cell (blue) with electrodes around it in a particular arrangement to generate a vertical field with a radial gradient for evaporative cooling. (**b**) shows a rotational transition moving by 10s of kHz after only 5 shots. (**c**) shows how the transition frequency relaxes over 10s of kHz over many hours. (**d**) shows how even the STIRAP resonance is shifted by the electric field induced by the charges on the glass, and shows how this induced field decays to zero over a timescale of several hours

high resolution optical detection. Moreover, the lack of an anti-reflection coating for the optical lattice beams made the vertical lattice path significantly more difficult to align in a way that minimized super-lattice effects (see Steven Moses' thesis [37]). The combination of this low quality glass cell and the limited optical access served as another source of motivation for the new apparatus.

While electric fields were used only minimally in published results from this experiment, we have spent a lot of time working with them over the years. Electric fields were generated by placing a small phenolic form between the inside of the IP trap (shown in Fig. 2.7) and the pyrex cell (shown in Fig. 2.8a). This phenolic form held indium-tin-oxide (ITO) coated glass plates, to which a small piece of copper foil was glued with a conductive epoxy. Then a high-voltage cable was soldered to the copper foil. The magnitude of the field was limited to 5–6 kV/cm by the nearby grounded coils on the IP trap (although we attempted to go higher in the earlier years), from which we were able to reach a dipole moment of 0.2 Debye [10, 18, 41].

The problem with electric fields that really limited us in the old chamber, however, was enormous transients and instability. We believe this is due to patch charges accumulated the glass between the electrodes, and even when the electric field was ostensibly zero, rotational transitions and even the STIRAP lineshapes were completely shifted by the electric field from the residual charges on the glass, as shown in Fig. 2.8b, c. Figure 2.8d shows how the field would relax on a timescale of roughly 5 h. This problem made it very difficult to do any rotational state spectroscopy in an electric field, and even made evaporative cooling of molecules prohibitively difficult. Figure 2.8 was also shown in Steven Moses' thesis because this was an extremely important problem which took a lot of our time in the early years.

Indeed, we tried many approaches to remove these charges, all of which were unsuccessful. We tried baking the chamber and shining UV lamps on it; we tried "de-Gaussing" the chamber with high voltage electrodes that we switched at various frequencies and pointed in various directions; and we even removed and replaced the glass cell on the chamber! This last attempt was actually a very large project, and required baking the chamber again to reach the typical vacuum lifetime. This latter effort actually worked, and all the charges were removed... until we applied large fields again and recharged the new pyrex cell, which was identical to the old cell. At this point we gave up on using electric fields for the next several years, and these problems served as the primary motivation for the second generation apparatus to be described later.

2.3 Moving the Experiment into the JILA X-Wing

At the end of 2012 and into the new year of 2013 we moved our experiment into the lab space that was prepared for us in the new X-wing of JILA. The two labs were only separated by 20–30 m, but there was a half-flight level change between the two wings, which required a crane as shown in Fig. 2.9. We were very careful during the process, and we closed the gate valve between the two halves of the chamber before

Fig. 2.9 Moving the lab. (**a**) shows the old lab, S1B05. (**b**) shows moving the experiment down a half-flight of stairs. (**c**) shows the new lab, X1B20, before moving the optical table from the other lab. Note that a second table is in place to the right, and this is the table where almost all of the lasers are situated

the move. Nothing went wrong during the move, and as with the OH experiment which had moved earlier to the other side of our new lab, our first paper in the X-wing was published in Nature. We were able to create a MOT within 1 week and a BEC within 2 months despite changing several aspects of the system.

2.4 Laser Systems in the X-Wing Lab

Here I will give a brief overview of all the laser systems used in the X-wing lab since 2012, which include the MOT/probe lasers, the Raman lasers, and the high power optical trapping and lattice lasers at 1064 nm. More recently we have added an MSquared tunable Ti:Sapphire laser, which I will discuss more during the chapter on the new apparatus.

2.4.1 MOT/Probe Lasers

While most of the experimental apparatus did not change as a result of the move, we took that opportunity to completely change the K and Rb trapping and probing lasers. The old systems were based on homemade external cavity diode lasers (ECDLs) and Newport Vortex lasers, which had become very unreliable and required frequent relocking. The new laser systems were built by Steven Moses, and they are based on distributed Bragg reflector (DBR) lasers from Photodigm. These lasers have high power ($\sim$100 mW), and very large mode-hop-free tuning ranges of 10s to 100s of GHz. Therefore, these lasers routinely stay locked for weeks. Eagleyard tapered amplifiers (TAs) are used to deliver enough power for the MOTs, which require $\approx$100–300 mW.

The repump lasers for both K and Rb are locked to an absorption cell using saturated absorption frequency-modulation (FM) spectroscopy. This is done using counter-propagating beams of a few 100 μW whose phase is modulated with an electro-optic modulator (EOM) at 20 MHz. The error signal is derived by demodulating the probe beam transmission with a phase-shifted rf tone of the modulation frequency [4].

The trap lasers are locked to their respective repump lasers by beating them on a fast photodiode and locking to a frequency offset. This is done by mixing the $\sim$GHz beat notes with a voltage controlled oscillator (VCO), and in the case of Rb we divide the beat note frequency by 8 before mixing with a VCO. After mixing these frequencies down, they are converted to a voltage with an F-V converter, which has an adjustable offset. While this is a very robust way to offset lock a laser, we are constantly limited by the stability of the VCO, which has a large thermal coefficient. Therefore, we recently updated our offset lock approach to use direct digital synthesis (DDS)-generated frequencies instead of VCOs.

2.4.2 Raman Lasers and Cavity

In 2012 Steven built a bichromatic high-finesse optical cavity for locking the Raman lasers, instead of the Ti:Sapphire comb that had been in use since 2007 or so. While the Ti:Sapphire system allowed for enormous tunability of the Raman lasers, which was critical during the initial spectroscopy period in 2007 and 2008, the stability of the Raman laser system was limited by the stability of the comb. During the first year of my PhD when we were locking the Raman lasers to the comb, it would take hours or even the full day in order to get the comb to stay robustly locked. The situation was further complicated by the fact that the comb was in a different room from the KRb experiment, with a long fiber strung between them.

We typically operate with 200 mW of power in the up leg and 20 mW in the down leg, whose transition dipole moments are 0.005(2) ea_0 and 0.012(3) ea_0, respectively [41]. Such powers are needed to achieve Rabi frequencies of a few MHz. For the up leg at 968 nm we use an Eagleyard anti-reflection (AR)-coated 980 nm laser diode in a Littrow configuration (the AR coating generally shifts the wavelength blue by ∼10 nm). This is then amplified with an Eagleyard 970 nm TA. The down leg is also an Eagleyard anti-reflection AR-coated diode laser operating at 689 nm in the Littrow configuration, though it has been changed several times over the years and was a Sacher Lasertechnik diode in the past. For the down leg we amplify the transmitted power through the cavity (∼500 µW) instead of the raw diode because it is much more stable. In the end we did not observe any significant effect on the STIRAP efficiency compared to using the down leg before the cavity, but nevertheless we inject an OpNext slave laser (HL6750MG) with the cavity-transmitted light, which is ∼100s of µW.

The cavity has a finesse of $\sim 2 \times 10^4$, and is comprised of a cylindrical piece of Zerodur with a bore through the center (see Fig. 2.10). The mirror substrates are fused silica; one mirror is flat and the other has a radius of curvature of 50 cm to ensure that the transverse modes are non-degenerate. The coating was done by Advanced Thin Films in Boulder, and the cavity is temperature stabilized to

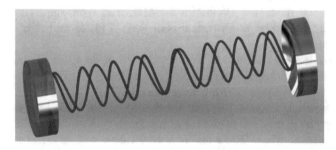

Fig. 2.10 The bichromatic optical cavity used to lock the Raman lasers. The two laser wavelengths are 690 nm (blue) and 970 nm (red). As shown in this figure, an integer number of wavelengths for both lasers is required to match the cavity length

$\sim 10^{-2}$ °C, from which its resonances drift by roughly 100 kHz daily. The two-photon STIRAP lineshape has a width of 500 kHz, so this drift can easily be corrected by optimizing the STIRAP efficiency every few days.

The lasers are locked to the cavity using a standard PDH locking technique [22] to the transverse electromagnetic TEM$_{0,0}$ mode. The servo bandwidths are ~ 2 MHz for both lasers. The PDH error signal, cavity-ringdown spectroscopy, and the cavity alignment are shown in Fig. 2.11. The frequency difference between the nearest TEM$_{0,0}$ cavity mode and the desired laser frequency is compensated using an AOM. Indeed, the STIRAP lineshape is scanned by adjusting the up leg AOM which is in a double-pass configuration. Thus, the down leg AOM is configured so that STIRAP operates on resonance with the intermediate state, as discussed earlier in this chapter, and the up leg double-pass AOM is adjusted to scan the two-photon resonance.

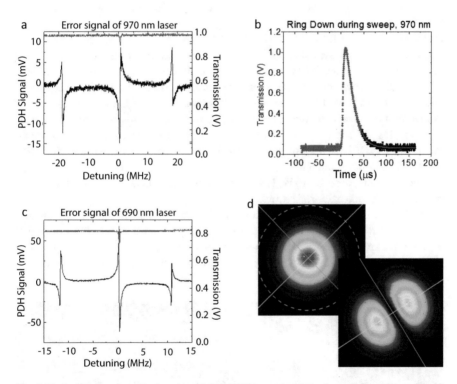

Fig. 2.11 Locking to the optical cavity. (**a**) The PDH error signal for the up leg at 970 nm (black) as a function of the detuning from the cavity resonance. The cavity transmission is also shown in red. (**b**) A cavity ring-down measurement with the up-leg, which is used to measure the finesse and the free-spectral range of the cavity. (**c**) The PDH error signal for the down leg at 690 nm (black) as a function of the detuning from the cavity resonance. The cavity transmission is also shown in red. Note that the transmission on resonance is very noisy. This measurement was taken with the diode at the end of its life, and was subsequently replaced. (**d**) The alignment procedure of the cavity produces many beautiful Hermite-Gaussian spatial modes. These images show the transverse electromagnetic TEM$_{0,0}$ and the TEM$_{1,0}$ modes

This Raman laser system has become incredibly robust, and is now one of the most reliable parts of the experiment.

2.4.3 Optical Trap/Lattice Lasers

All the optical traps and lasers are at $\lambda = 1064$ nm, and while the laser systems have changed many times over the years, we have been using a 50 W Nufern fiber amplifier since 2013 when we moved to the X-wing. This amplifier is seeded with a Mephisto NPRO (non-planar ring oscillator) solid-state laser, which has a linewidth of $\sim$10s of kHz. The linewidth of the Nufern after amplification is increased to $\sim$100s of kHz, which is still quite good. Mephisto now makes their own laser amplifier to accompany their NPRO which has an unequivocally lower noise profile than the Nufern, and most optical lattice groups have moved away from Nufern lasers. However, we have been using the Nufern since before the Mephisto option was popular or even available, and we believe that we are not as sensitive to laser noise on our lattices as, for example, lithium quantum gas microscope experiments.

All of our beams are derived from the Nufern, and so the power needs to be intelligently distributed between many paths in a way that maximizes the power available at any stage of the experiment. To accomplish this we use a layout as shown in Fig. 2.12, where the power follows a path through several acousto-optic modulators (AOMs), and can be diverted to each fiber at will using each AOM. Typical AOM diffraction and fiber coupling efficiencies are $\sim$80%, and we typically operate at a total power of 40 W. Note that recently an isolator has been added to the right arm to reduce back-reflection.

While this layout style is inherently challenging to maintain since any change to an AOM will severely affect the subsequent paths, we eventually converged on a setup that is quite robust. We use a photodiode to detect the power going to either arm, as controlled with the waveplate before the polarizing beam splitter at

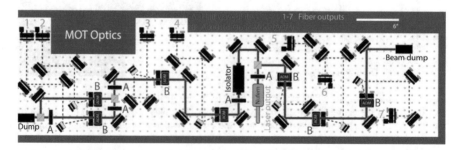

Fig. 2.12 The Nufern fiber amplifier layout. The main beam path "arms" are shown in thick red, and diffraction from each AOM is shown in dotted red. Obstructions from other optics and other parts of the experiment are shown in dark grey. Note that an isolator has since been added to the right arm

the beginning of the path. The photodiodes are placed at the end of each arm, and are picked off from the beam before the water-cooled beam dumps using back-side polished mirrors. These mirrors reflect ∼99.5% of the light, so looking at the transmitted portion is inherently noisy since a 0.1% change in the reflection can be a 100% change in the transmission. Indeed, we see ∼30% fluctuations in the photodiode voltages over several days.

These photodiodes are very important for us because we see enormous polarization fluctuations coming out of the Nufern. We spent a large amount of time trying to understand and remove these fluctuations, but they appear to be caused by the laser mode coming out of the fiber from the amplifier. To solve this problem, we use a Thorlabs motorized rotation mount with a $\lambda/2$ waveplate, which we feed back to using one of the photodiode voltages and regulate with a digital servo. This works very well, although we occasionally have to change the servo setpoint slightly as the power fraction to the photodiode drifts.

Another major complication from the Nufern is its inherent pointing instability. We see that the direction the beam is coming out of the fiber changes with time, and typically has a ≈100 μm RMS deviation over a distance of ≈1 m, over timescales of 10s of minutes to hours. We think that this issue also comes from the mode of the light in the fiber, which changes slightly in time. Therefore, this pointing fluctuation appears to be correlated with the polarization fluctuations. It is particularly problematic for us given the length and complexity of our two arms. We tried to use a piezo-actuated mirror and a quadrant photodiode to servo this pointing drift, but our ability to fully correct for this is limited by how close we can place the mirror to the Nufern fiber output. Thus, in the end we have to re-optimize most of the path alignment every few weeks.

Some of these issues are potentially exacerbated by the fact that we were using the JILA chilled water system to cool our Nufern, which operates at ≈17 °C. The output power of the Nufern is strongly dependent on the cooling temperature as shown in Fig. 2.13a, and the recommended operating temperature is 23 °C.

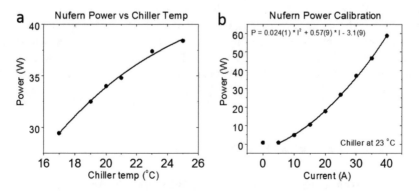

Fig. 2.13 The power calibration of the Nufern. (**a**) The power as a function of the chiller temperature. (**b**) The power as a function of current for a chiller temperature of 23 °C

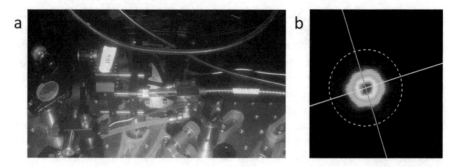

Fig. 2.14 The hollow-core photonic crystal fiber. (**a**) The brass cooling block attached to the fiber connector, and the coupling optics for the fiber. (**b**) The profile of a focused beam far from the fiber output. Note the hexagonal shape caused by the core structure of the fiber. Any aberrations wash out the corners

Accordingly, we switched from the JILA chilled water system to a high cooling power chiller which allows us to vary the temperature. With the laser operating at the optimal temperature we can measure the power as a function of the input current, as in Fig. 2.13b. Cooling the laser to its optimum temperature helps to mitigate the pointing and polarization drift, though the difference has not been carefully quantified.

For beams where more than 3 W is delivered to the atoms we use hollow core photonic crystal fibers from TraTech FiberOptics. We water cool the brass connector at the end of the jacket using a cooling block which also provides mechanical stress relief [44] (see Fig. 2.14a). The cooling plates have holes for two thermistors; one for monitoring the temperature, and the other for an interlock on the corresponding AOM rf power. We typically optimize the fiber alignment by pulsing the AOM with a 5% duty cycle. The temperature change is <0.5 °C when well aligned even with an input of 20 W. To couple light into and out of these fibers we use a plano-convex fused silica lens with focal length $f = 20$ mm, since the normal Thorlabs aspheres are made of KB-7 or other glasses with higher absorption. When the fiber output is focused with minimal aberrations, a slightly hexagonal shape is apparent even in the far field. The initial coupling of light into these fibers is very difficult, and we've gotten pretty good at coupling light from a fiber alignment pen through the other side of the fiber without disturbing its mounting, or any optics between it and the vacuum chamber.

2.5   Computer Control Software

Until very recently the computer control software for this experiment has never been updated from the original mid-2000s design. The control program was written in Labview in the early 2000s, and I think Debbie Jin and Scott Papp were heavily involved in its original construction. This program controls 64 digital output lines using a digital input/output (DIO) board, which is made by Viewpoint Systems (64-DIO64PCI). The board's internal clock sets the timing for the entire experimental sequence. There are also three digital to analog converter (DAC) boards made by National Instruments (NI-PCI6733), which are synchronized with the DIO board. Two of the DAC boards generate 16 analog outputs, and the other is used to generate the detailed time-dependent ramps for optical traps. We call the former "static DACs" and the latter "dynamics DACs." The static DACs can perform linear ramps, but cannot be changed dynamically and cannot do complicated waveforms.

In recent years we have run out of both digital and analog lines, and so we have begun to add many FPGA and DDS boards that operate in parallel with the Labview architecture. These boards are controlled with Matlab and Labview. Moreover, before we implemented the new apparatus chamber, we attempted to update our DAC boards and the version of Labview used to control the experiment. We run with Labview 7.1 which requires Windows XP. We bought new DAC boards (identical boards) to replace the old ones, but these also require a newer version of Labview and Windows. In the end we found it to be very complicated to update the version of Labview to run on Windows 7 or Windows 10 while maintaining the original Labview program. Therefore, until recently we have been operating with a very old computer control system.

We have become increasingly aware of problems with this system, as well as residual timing-based bugs in the Labview control program. Over the years many digital lines have begun to fail, and some of the DAC channels are also unreliable. Recently we learned that the DACs are unstable at the 10s of mV level such that ramps are jagged and non-monotonic. Therefore, once we created a degenerate K-Rb mixture and KRb molecules in the new apparatus, we decided to update our computer control architecture.

We recently switched to an FPGA-based control software which is controlled with Python. This program was recently designed by the Strontium group in Jun's lab, and Ross Hutson in particular. Both the analog and digital controls are derived from FPGAs, such that eight are used in total. Two boards with 64 digital lines are used to generate 128 digital lines, and six boards with 8 DACs are used for 64 analog DAC lines.

References

1. K. Aikawa, D. Akamatsu, J. Kobayashi, M. Ueda, T. Kishimoto, S. Inouye, Toward the production of quantum degenerate bosonic polar molecules, 41 k 87 rb. New J. Phys. **11**(5), 055035 (2009)
2. M.H. Anderson, J.R. Ensher, M.R. Matthews, C.E. Wieman, E.A. Cornell, Observation of Bose-Einstein condensation in a dilute atomic vapor. Science **269**(5221), 198–201 (1995)
3. M. Bartenstein, A. Altmeyer, S. Riedl, S. Jochim, C. Chin, J. Hecker Denschlag, R. Grimm, Crossover from a molecular bose-einstein condensate to a degenerate fermi gas. Phys. Rev. Lett. **92**, 120401 (2004)

4. G.C. Bjorklund, Frequency-modulation spectroscopy: a new method for measuring weak absorptions and dispersions. Opt. Lett. **5**(1), 15–17 (1980)

5. C.C. Bradley, C.A. Sackett, J.J. Tollett, R.G. Hulet, Evidence of Bose-Einstein condensation in an atomic gas with attractive interactions. Phys. Rev. Lett. **75**, 1687–1690 (1995)

6. C.C. Bradley, C.A. Sackett, R.G. Hulet, Bose-Einstein condensation of lithium: observation of limited condensate number. Phys. Rev. Lett. **78**, 985–989 (1997)

7. J.M. Brown, A. Carrington, *Rotational Spectroscopy of Diatomic Molecules* (Cambridge University Press, Cambridge, 2003)

8. L.D. Carr, D. DeMille, R.V. Krems, J. Ye, Cold and ultracold molecules: science, technology and applications. New J. Phys. **11**(5), 055049 (2009)

9. C. Chin, R. Grimm, P. Julienne, E. Tiesinga, Feshbach resonances in ultracold gases. Rev. Mod. Phys. **82**, 1225–1286 (2010)

10. A. Chotia, B. Neyenhuis, S.A. Moses, B. Yan, J.P. Covey, M. Foss-Feig, A.M. Rey, D.S. Jin, J. Ye, Long-lived dipolar molecules and feshbach molecules in a 3D optical lattice. Phys. Rev. Lett. **108**, 080405 (2012)

11. E.U. Condon, Nuclear motions associated with electron transitions in diatomic molecules. Phys. Rev. **32**, 858–872 (1928)

12. J.P. Covey, S.A. Moses, M. Garttner, A. Safavi-Naini, M.T. Miecnkowski, Z. Fu, J. Schachenmayer, P.S. Julienne, A.M. Rey, D.S. Jin, J. Ye, Doublon dynamics and polar molecule production in an optical lattice. Nat. Commun. **7**, 11279 (2016)

13. J.G. Danzl, E. Haller, M. Gustavsson, M.J. Mark, R. Hart, N. Bouloufa, O. Dulieu, H. Ritsch, H.-C. Nägerl, Quantum gas of deeply bound ground state molecules. Science **321**(5892), 1062–1066 (2008)

14. J.G. Danzl, M.J. Mark, E. Haller, M. Gustavsson, R. Hart, A. Liem, Z. Holger, H.-C. Nägerl, Deeply bound ultracold molecules in an optical lattice. New J. Phys. **11**(5), 055036 (2009)

15. J.G. Danzl, M.J. Mark, E. Haller, M. Gustavsson, R. Hart, J. Aldegunde, J.M. Hutson, H.-C. Nägerl, An ultracold high-density sample of rovibronic ground-state molecules in an optical lattice. Nat. Phys. **6**, 265–270 (2010)

16. K.B. Davis, M.O. Mewes, M.R. Andrews, N.J. van Druten, D.S. Durfee, D.M. Kurn, W. Ketterle, Bose-Einstein condensation in a gas of sodium atoms. Phys. Rev. Lett. **75**, 3969–3973 (1995)

17. M.H.G. de Miranda, Control of dipolar collisions in the quantum regime. PhD thesis, University of Colorado, Boulder, 2010

18. M.H.G. de Miranda, A. Chotia, B. Neyenhuis, D. Wang, G. Quéméner, S. Ospelkaus, J.L. Bohn, J. Ye, D.S. Jin, Controlling the quantum stereodynamics of ultracold bimolecular reactions. Nat. Phys. **7**(6), 502–507 (2011)

19. B. DeMarco, D.S. Jin, Onset of fermi degeneracy in a trapped atomic gas. Science **285**(5434), 1703–1706 (1999)

20. K. Dieckmann, C.A. Stan, S. Gupta, Z. Hadzibabic, C.H. Schunck, W. Ketterle, Decay of an ultracold fermionic lithium gas near a Feshbach resonance. Phys. Rev. Lett. **89**, 203201 (2002)

21. E.A. Donley, N.R. Claussen, S.T. Thompson, C.E. Wieman, Atom-molecule coherence in a bose-einstein condensate. Nature **417**, 529 (2002)

22. R.W.P. Drever, J.L. Hall, F.V. Kowalski, J. Hough, G.M. Ford, A.J. Munley, H. Ward, Laser phase and frequency stabilization using an optical resonator. Appl. Phys. B **31**(2), 97–105 (1983)

23. J. Franck, E.G. Dymond, Elementary processes of photochemical reactions. Trans. Faraday Soc. **21**, 536–542 (1926)

24. P.D. Gregory, P.K. Molony, M.P. Köppinger, A. Kumar, Z. Ji, B. Lu, A.L. Marchant, S.L. Cornish, A simple, versatile laser system for the creation of ultracold ground state molecules. New J. Phys. **17**(5), 055006 (2015)

25. M. Greiner, C.A. Regal, D.S. Jin, Emergence of a molecular Bose-Einstein condensate from a Fermi gas. Nature **426**, 537–540 (2003)

26. M. Guo, B. Zhu, B. Lu, X. Ye, F. Wang, R. Vexiau, N. Bouloufa-Maafa, G. Quéméner, O. Dulieu, D. Wang, Creation of an ultracold gas of ground-state dipolar $^{23}Na^{87}Rb$ molecules. Phys. Rev. Lett. **116**, 205303 (2016)

27. J. Herbig, T. Kraemer, M. Mark, T. Weber, C. Chin, H.-C. Nägerl, R. Grimm, Preparation of a pure molecular quantum gas. Science **301**(5639), 1510–1513 (2003)

28. E.R. Hudson, N.B. Gilfoy, S. Kotochigova, J.M. Sage, D. DeMille, Inelastic collisions of ultracold heteronuclear molecules in an optical trap. Phys. Rev. Lett. **100**, 203201 (2008)

29. S. Jochim, M. Bartenstein, A. Altmeyer, G. Hendl, S. Riedl, C. Chin, J. Hecker Denschlag, R. Grimm, Bose-Einstein condensation of molecules. Science **302**, 2101–2103 (2003)

30. K.M. Jones, E. Tiesinga, P.D. Lett, P.S. Julienne, Ultracold photoassociation spectroscopy: long-range molecules and atomic scattering. Rev. Mod. Phys. **78**, 483–535 (2006)

31. M. Knap, E. Berg, M. Ganahl, E. Demler, Clustered Wigner-crystal phases of cold polar molecules in arrays of one-dimensional tubes. Phys. Rev. B **86**, 064501 (2012)

32. T. Köhler, K. Góral, P.S. Julienne, Production of cold molecules via magnetically tunable Feshbach resonances. Rev. Mod. Phys. **78**, 1311–1361 (2006)

33. K.A. Kuns, A.M. Rey, A.V. Gorshkov, d-Wave superfluidity in optical lattices of ultracold polar molecules. Phys. Rev. A **84**, 063639 (2011)

34. F. Lang, K. Winkler, C. Strauss, R. Grimm, J. Hecker Denschlag, Ultracold triplet molecules in the rovibrational ground state. Phys. Rev. Lett. **101**, 133005 (2008)

35. P. Marte, P. Zoller, J.L. Hall, Coherent atomic mirrors and beam splitters by adiabatic passage in multilevel systems. Phys. Rev. A **44**, R4118–R4121 (1991)

36. P.K. Molony, P.D. Gregory, Z. Ji, B. Lu, M.P. Köppinger, C.R. Le Sueur, C.L. Blackley, J.M. Hutson, S.L. Cornish, Creation of ultracold $^{87}Rb^{133}Cs$ molecules in the rovibrational ground state. Phys. Rev. Lett. **113**, 255301 (2014)

37. S.A. Moses, A quantum gas of polar molecules in an optical lattice. PhD thesis, University of Colorado, Boulder, 2016

38. S.A. Moses, J.P. Covey, M.T. Miecnikowski, B. Yan, B. Gadway, J. Ye, D.S. Jin, Creation of a low-entropy quantum gas of polar molecules in an optical lattice. Science **350**(6261), 659–662 (2015)

39. B. Neyenhuis, Ultracold polar krb molecules in optical lattices. PhD thesis, University of Colorado, Boulder, 2012

40. K.-K. Ni, A quantum gas of polar molecules. PhD thesis, University of Colorado, Boulder, 2009

41. K.-K. Ni, S. Ospelkaus, M.H.G. de Miranda, A. Pe'er, B. Neyenhuis, J.J. Zirbel, S. Kotochigova, P.S. Julienne, D.S. Jin, J. Ye, A high phase-space-density gas of polar molecules. Science **322**(5899), 231–235 (2008)

42. M.L. Olsen, J.D. Perreault, T.D. Cumby, D.S. Jin, Coherent atom-molecule oscillations in a Bose-Fermi mixture. Phys. Rev. A **80**, 030701 (2009)

43. C.H.R. Ooi, Laser cooling of molecules by zero-velocity selection and single spontaneous emission. Phys. Rev. A **82**, 053408 (2010)

44. N.D. Oppong, Towards a degenerate fermi gas of strontium-87 in a 3d optical lattice. MS thesis, University of Colorado, Boulder, 2015

45. S. Ospelkaus, A. Pe'er, K.-K. Ni, J.J. Zirbel, B. Neyenhuis, S. Kotochigova, P.S. Julienne, J. Ye, D.S. Jin, Efficient state transfer in an ultracold dense gas of heteronuclear molecules. Nat. Phys. **4**, 622–626 (2008)

46. J.W. Park, S. Will, M.W. Zwierlein, Ultracold dipolar gas of fermionic $^{23}Na^{40}K$ molecules in their absolute ground state. Phys. Rev. Lett. **114**, 205302 (2015)

47. G. Reinaudi, C.B. Osborn, M. McDonald, S. Kotochigova, T. Zelevinsky, Optical production of stable ultracold $^{88}Sr_2$ molecules. Phys. Rev. Lett. **109**, 115303 (2012)

48. A. Safavi-Naini, M.L. Wall, A.M. Rey, Role of interspecies interactions in the preparation of a low-entropy gas of polar molecules in a lattice. Phys. Rev. A **92**, 063416 (2015)

49. S. Stellmer, B. Pasquiou, R. Grimm, F. Schreck, Creation of ultracold Sr_2 molecules in the electronic ground state. Phys. Rev. Lett. **109**, 115302 (2012)

50. N. Syassen, D.M. Bauer, M. Lettner, T. Volz, D. Dietze, J.J. García-Ripoll, J.I. Cirac, G. Rempe, S. Dürr, Strong dissipation inhibits losses and induces correlations in cold molecular gases. Science **320**(5881), 1329–1331 (2008)

51. S.V. Syzranov, M.L. Wall, V. Gurarie, A.M. Rey, Spin–orbital dynamics in a system of polar molecules. Nat. Commun. **5**, 5391 (2014)

52. T. Takekoshi, L. Reichsöllner, A. Schindewolf, J.M. Hutson, C.R. Le Sueur, O. Dulieu, F. Ferlaino, R. Grimm, H.-C. Nägerl, Ultracold dense samples of dipolar RbCs molecules in the rovibrational and hyperfine ground state. Phys. Rev. Lett. **113**, 205301 (2014)

53. A.G. Truscott, K.E. Strecker, W.I. McAlexander, G.B. Partridge, R.G. Hulet, Observation of fermi pressure in a gas of trapped atoms. Science **291**(5513), 2570–2572 (2001)

54. R. Unanyan, M. Fleischhauer, B.W. Shore, K. Bergmann, Robust creation and phase-sensitive probing of superposition states via stimulated Raman adiabatic passage (STIRAP) with degenerate dark states. Opt. Commun. **155**(13), 144–154 (1998)

55. D. Wang, B. Neyenhuis, M.H.G. de Miranda, K.-K. Ni, S. Ospelkaus, D.S. Jin, J. Ye, Direct absorption imaging of ultracold polar molecules. Phys. Rev. A **81**, 061404 (2010)

56. K. Winkler, F. Lang, G. Thalhammer, P. von der Straten, R. Grimm, J.H. Denschlag, Coherent optical transfer of Feshbach molecules to a lower vibrational state. Phys. Rev. Lett. **98**, 043201 (2007)

57. C.-H. Wu, J.W. Park, P. Ahmadi, S. Will, M.W. Zwierlein, Ultracold fermionic feshbach molecules of $^{23}Na^{40}K$. Phys. Rev. Lett. **109**, 085301 (2012)

58. N.Y. Yao, A.V. Gorshkov, C.R. Laumann, A.M. Läuchli, J. Ye, M.D. Lukin, Realizing fractional chern insulators in dipolar spin systems. Phys. Rev. Lett. **110**, 185302 (2013)

59. L.P. Yatsenko, B.W. Shore, K. Bergmann, Detrimental consequences of small rapid laser fluctuations on stimulated Raman adiabatic passage. Phys. Rev. A **89**, 013831 (2014)

60. J.J. Zirbel, Ultracold fermionic feshbach molecules. PhD thesis, University of Colorado, Boulder, 2008

61. M.W. Zwierlein, C.A. Stan, C.H. Schunck, S.M.F. Raupach, S. Gupta, Z. Hadzibabic, W. Ketterle, Observation of Bose-Einstein condensation of molecules. Phys. Rev. Lett. **91**, 250401 (2003)

Chapter 3
Quantum-State Controlled Chemical Reactions and Dipolar Collisions

Immediately after ultracold polar KRb molecules were produced in the rovibrational ground state, as well as in the lowest hyperfine state, it became clear that chemical reactions were responsible for the loss of molecules confined in a conventional far-off-resonance optical dipole trap [18, 20], as the reaction $2KRb \rightarrow K_2 + Rb_2$ is exothermic. For understanding the nature of the dipolar collisions between molecules in the quantum regime, it is imperative to have a thorough understanding of the molecular states involved and the resulting quantum statistics of the colliding molecules. In this chapter I will describe the experiments that led up to the beginning of my PhD research, which is discussed in the next chapter. These discussions can also be found in our book chapter in "Low energy and low temperature molecular scattering" [5].

3.1 Quantum State Control of the Molecule

STIRAP transfers the molecules to the rovibrational ground state of the singlet $(X^1\Sigma^+)$ electronic ground potential. Although the electronic angular momentum is zero, there is hyperfine structure arising from the nuclear magnetic moments of the constituent atoms. K has nuclear quantum number $I^K = 4$, while Rb has nuclear quantum number $I^{Rb} = 3/2$, so there are 36 hyperfine states for each rotational state $|N, m_N\rangle$, where N is the principal rotational quantum number of KRb and m_N is its projection onto the quantization axis (which can be defined by either an electric or a magnetic field applied in the lab). These states can be labeled as $|N, m_N, m_K, m_{Rb}\rangle$, where m_K and m_{Rb} are the nuclear spin projection quantum number for K and Rb, respectively. Note that since the ground electronic potential of KRb has both the orbital and spin angular momentum equal to zero,

© Springer Nature Switzerland AG 2018   31
J. P. Covey, *Enhanced Optical and Electric Manipulation of a Quantum Gas of KRb Molecules*, Springer Theses, https://doi.org/10.1007/978-3-319-98107-9_3

the rotations of the molecular nuclei, **N**, are decoupled from the nuclear spin, **I**. Therefore, $|N, m_N, m_K, m_{Rb}\rangle$ is an appropriate set of quantum numbers for such molecules [2]. In the ultracold regime, the initial Feshbach association creates weakly-bound molecules in a single quantum state since K and Rb atoms approach each other in the lowest partial wave, s ($L = 0$), where $\hbar\mathbf{L}$ is the quantized relative angular momentum. Thus, through a combination of angular momentum selection rules and the spectral resolution provided by STIRAP, the ground-state molecules also occupy a single hyperfine quantum state, which for KRb is $|0, 0, -4, 1/2\rangle$. The lowest energy hyperfine state is $|0, 0, -4, 3/2\rangle$ [19]. To reach this absolute ground state and to also provide a general spectroscopic tool to coherently and efficiently transfer populations between different hyperfine states, we have developed another coherent two-photon transfer technique, which is based on the use of two microwave photons to couple from two different hyperfine states in $|N = 0\rangle$ to a common hyperfine state in the $|N = 1\rangle$ manifold via electric dipole transitions. This two-photon coupling is enabled by the fact that the $|N = 1\rangle$ state has non-negligible (a few percent) spin mixing due to coupling between rotation and the nuclear electric quadrupole moment [1, 2, 19]. This microwave-based Raman transition allows us to transfer the molecular population to any hyperfine state in $|N = 0\rangle$, including the absolute ground state $|0, 0, -4, 3/2\rangle$. This technique is also important for the creation of a 50:50 mixture of two hyperfine states in the ground state, or a 50:50 mixture of two rotational states for KRb molecules, which we use to study the role of quantum statistics of identical molecules in the reaction process.

3.2 Inducing the Dipole Moment in a Lab Frame with DC and AC Electric Fields

Since an electric dipole moment exists only between quantum states that have opposite parity, individual rotational states $|N, m_N\rangle$ have no dipole moment in the laboratory frame. In other words, molecules are polarized along their internuclear axis, but since this axis changes rapidly and randomly in the absence of an applied electric field, the dipole moment averages to zero in the laboratory frame. When a DC electric field is applied, opposite-parity rotational states become mixed, giving rise to a nonzero dipole moment in the lab frame. In other words, molecules become aligned with the applied field. The Hamiltonian for the rotational states $|N, m_N\rangle$ under an electric field (ϵ) has two contributions arising from the diagonal rotational energy and the state-mixing (off-diagonal) Stark effect:

$$\langle N, m_N|\hat{H}|N', m'_N\rangle = BN(N+1)\delta_{N,N'}\delta_{m_N,m_{N'}} - \langle N, m_N|\hat{\mathbf{d}}\cdot\epsilon|N', m'_N\rangle. \quad (3.1)$$

Here $B = \hbar^2/2I$ is the rotational constant for the state under consideration (I is the moment of inertia, which is the reduced mass ϕ times the square of the internuclear separation), and the term $-\hat{\mathbf{d}} \cdot \epsilon$ describes the dipole-field coupling operator. If the electric field, of magnitude ϵ, is defined along the z axis (indeed the electric field defines the quantization axis), the matrix elements of Eq. (3.1) simplify to

$$\langle N, m_N | \hat{H} | N', m'_N \rangle = BN(N+1)\delta_{N,N'}\delta_{m_N,m_{N'}} -$$

$$\mathcal{D}\epsilon\sqrt{(2N+1)(2N'+1)}(-1)^{m_N} \begin{pmatrix} N & 1 & N' \\ -m_N & 0 & m'_N \end{pmatrix} \begin{pmatrix} N & 1 & N' \\ 0 & 0 & 0 \end{pmatrix}. \quad (3.2)$$

Here $\mathcal{D}$ is the permanent, molecule frame dipole matrix element (0.574 Debye for KRb [17]) and the terms in parentheses are Wigner 3-j symbols [8]. The first 3-j symbol is zero unless $m_N = m_{N'}$. So the eigenstates in an electric field are

$$|\tilde{N}, m_N \rangle = \sum_{N'} c_{N'} |N', m_N \rangle, \quad (3.3)$$

where $|\tilde{N}, m_N \rangle$ is the state that adiabatically connects to $|N, m_N \rangle$ when the field is turned down to zero. The dressed state energies $E_{\tilde{N},m_N}$ can be determined by diagonalizing Eq. (3.2), and then the induced dipole moment of state $|\tilde{N}, m_N \rangle$ is known through the quantity, $-\frac{\partial E_{\tilde{N},m_N}}{\partial \epsilon}$. The induced dipole moments for the $N = 0$ and $N = 1$ states of KRb are displayed in Fig. 3.1. A field strength of more than 40 kV/cm in the lab is required to reach 80% of the intrinsic dipole moment of the molecules. In general, the value of this critical field scales as $B/\mathcal{D}$, and it can be much smaller for some bialkali molecular species.

Since successive rotational states have opposite parity, a more efficient option for realizing a lab-frame dipole moment is to use a resonant rf or microwave field to directly couple two neighboring rotational states. When the $N = 0$ and $N = 1$ states are coherently mixed by applying a microwave pulse that creates a superposition (e.g., a $\pi/2$-pulse), we realize a dipole moment of magnitude $\mathcal{D}/\sqrt{3}$ in the rotating frame of the molecule. This is the technique used in Refs. [11, 24] to realize dipolar-interaction-mediated spin exchanges in a three-dimensional optical lattice, to be discussed in Chap. 5.

Fig. 3.1 Induced dipole moments of KRb for the $|\tilde{0}, 0\rangle$ (blue), $|\tilde{1}, 0\rangle$ (orange), and $|\tilde{1}, \pm 1\rangle$ states (green), calculated using the lowest 20 rotational levels in Eq. (3.2)

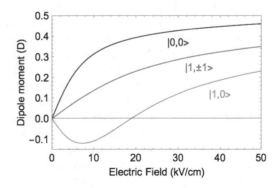

3.3   Role of Quantum Statistics on Collisions

Owing to the antisymmetrization requirement of the overall wavefunction for two identical fermions, spin-polarized (i.e., the same internal state) $^{40}K^{87}Rb$ molecules collide via odd partial waves. At ultralow temperatures, the lowest odd partial wave is the p-wave ($L = 1$). In contrast, distinguishable molecules (such as a hyperfine or a rotational mixture) can collide via s-wave ($L = 0$). Thus, when molecules are brought to the quantum regime, their interactions and reactions will be governed by the molecular quantum statistics, single partial-wave scattering, and quantum threshold laws at a vanishing collision energy. The first set of experiments that explored molecular reactions is reported in Ref. [20], which studied the dependence of reaction rate on temperature and quantum statistics without applying a laboratory electric field.

As a result of the exothermic exchange chemical reaction ($2KRb \rightarrow K_2 + Rb_2$), the number density n of trapped molecules decreases according to

$$\frac{dn}{dt} = -\beta n^2 - \alpha n, \tag{3.4}$$

where β is the two-body loss coefficient and α is the single-particle loss coefficient, which is set by a combination of effects including heating of the gas in the trap, light scattering from the trapping beams, and collisions with background particles. The origin of this heating is associated with the enhanced loss at the trap center where the density is the highest [20]. The remaining one-body loss mechanism has a much longer time constant. The two-body loss rate β is found to be proportional to T for spin-polarized samples, which is the expected dependence from the Bethe-Wigner threshold laws for p-wave collisions (Fig. 3.2). The lack of any non-monotonic behavior in the rate coefficient dependence on temperature is consistent with a simple model that assumes that after the molecules tunnel through the p-wave barrier they chemically react with a near unity probability [12].

For an incoherent 50:50 mixture of two hyperfine states in the ground state ($|0, 0, -4, 1/2\rangle$ and $|0, 0, -4, 3/2\rangle$), s-wave collisions start to play a dominant role since the molecules from the two different states are distinguishable. As a result, the loss rate shows no measurable temperature dependence and its magnitude is increased by ten- to hundredfold in comparison with the p-wave rate. The measured rate agrees well with the predicted universal loss rate related to the two-body van der Waals length. We note that the absolute ground state ($|0, 0, -4, 3/2\rangle$) demonstrates the same loss rate as other higher lying hyperfine states, which implies that the chemical reactions dominate over hyperfine state-changing collisions. Molecules prepared in an incoherent 50:50 mixture of rotational states (half in $|0, 0, -4, 1/2\rangle$ and half in $|1, -1, -4, 1/2\rangle$) have the largest loss rate observed at zero electric field (see Fig. 3.3). Finally, when both atoms and molecules are prepared in their lowest-energy states, we have also observed a universal s-wave loss rate from the exothermic atom-exchange reaction for $K + KRb \rightarrow K_2 + Rb$. For the endothermic process $Rb + KRb \rightarrow Rb_2 + K$, the observed loss rate approaches zero within

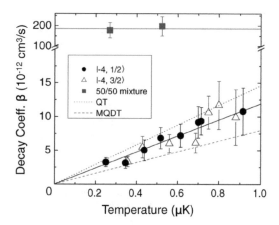

Fig. 3.2 Loss rate coefficient as a function of temperature in a bulk gas for spin-polarized $^{40}K^{87}Rb$ molecules, showing a linear increase with temperature characteristic of the p-wave threshold behavior. For comparison, loss rate for a 50:50 mixture of hyperfine spin states is shown, where the nature of s-wave collisions removes the temperature dependence for the rate constant while the loss magnitude is increased by a factor of 10–100. Reproduced from Ref. [20]

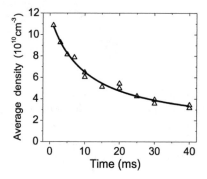

Fig. 3.3 Decay of an incoherent mixture of rotational states $|0, 0, -4, 1/2\rangle$ and $|1, -1, -4, 1/2\rangle$ of $^{40}K^{87}Rb$ molecules in a harmonic trap. The extracted loss rate is $9.0(4) \times 10^{-10} \, cm^3 \, s^{-1}$, which is almost 5 times higher than for the ground-state hyperfine mixture. Reproduced from Ref. [24]

the measurement uncertainty [20]. However, if the Rb density becomes too high, a three-body loss process involving two Rb atoms and a KRb will limit the molecular lifetime, as discussed in Chap. 6.

When an electric field is applied, the molecules develop an electric dipole moment in the laboratory frame and this modifies how they interact with each other. First, the isotropic centrifugal barrier arising from the p-wave becomes anisotropic. For $L = 1$, the projection onto the quantization axis (defined by the applied electric field), m_L, dictates how the molecules collide: $m_L = 0$ corresponds to a "head-to-tail" collision, which is attractive under the dipole–dipole interaction; while $m_L = \pm 1$ corresponds to a "side-by-side" collision, which is repulsive under

Fig. 3.4 Collision of two polar molecules in an electric field. (**a**) The p-wave barrier at zero electric field is isotropic and does not depend on the orientation in which the molecules collide. (**b**) In an applied DC electric field, dipolar interactions reduce the centrifugal barrier for "head-to-tail" collisions (blue arrow) and increase the centrifugal barrier for "side-by-side" collisions. Reproduced from Ref. [18]

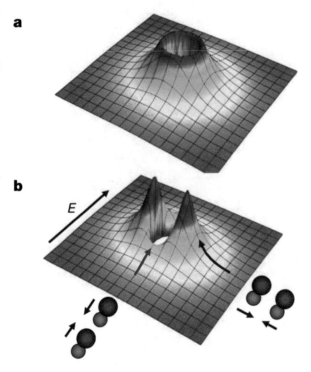

the dipole–dipole interaction. This is shown schematically in Fig. 3.4. Furthermore, the dipole–dipole interaction mixes states with different partial waves, such that the $L = 1$ partial wave is replaced by the lowest-energy channel with odd parity. Similarly, the $L = 0$ partial wave becomes the lowest-energy channel with even parity, which does not have a centrifugal barrier.

In a 3D optical dipole trap, we do not have control over the projection of the relative angular momentum of the colliding particles on the quantization axis. Therefore, while for $m_L = \pm 1$ the centrifugal barrier is significantly increased with a reasonable electric field, the observed loss rate is dominated by the $m_L = 0$ collision channel, which has a much lower barrier height for reactions (Fig. 3.4). Consequently, the measured loss rate increased sharply as a function of the electric field, as shown in Fig. 3.5. In fact, $\beta \propto d^6$ (d is the induced dipole moment in the laboratory frame), which agrees well with a quantum threshold model that takes into account the long-range dipolar interaction [18].

Chemical reactions for ultralow temperature polar molecules confined in a 3D harmonic trap thus depend strongly on the quantum statistics of the molecular gas, individual collision partial waves, temperature, and the induced dipole moment in the lab frame. In the remainder of this chapter I present results on the control of the reaction rate when we modify the dimensionality of the optical traps in which the molecules are confined.

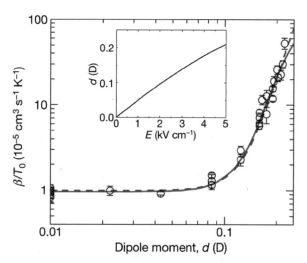

Fig. 3.5 Decay coefficient of ultracold $^{40}K^{87}Rb$ molecules colliding in an electric field as a function of induced dipole moment in a bulk 3D gas. A clear onset of increased loss at 0.1 D is apparent. The inset shows the induced dipole moment as a function of electric field. Reproduced from Ref. [18]

3.4   Reduced Dimensions: Quantum Stereodynamics of Chemical Reactions in a 2D Gas

To suppress the large reaction rate of KRb molecules in an electric field, we must suppress the head-to-tail collisions that lead to inelastic loss. Hence, we need to gain control over the projection of the relative angular momentum of the colliding molecules on the quantization axis, m_L. This can be achieved by confining molecules in a two-dimensional optical trap, which can be realized by interfering two counter-propagating optical beams to create a 1D lattice that consists of an array of 2D disk-shaped traps. To control the stereodynamics of the reaction (the orientation and relative motion of the reactants), a DC electric field is applied along the z axis and perpendicular to the 2D traps. In such a configuration, the lossy head-to-tail collisions with $m_L = 0$ are greatly suppressed, and the repulsive side-by-side collisions with $m_L = \pm1$ stabilize the molecular gas [3, 7, 15].

For sufficiently strong optical confinement along z, the motion along z is fully quantized and can be labeled with a quantum number ν_z (which is the band index, to be discussed in Chap. 4). At ultralow temperatures, the majority of molecules occupy the motional ground state $\nu_z = 0$, with only a small fraction of them in $\nu_z = 1$. The angular momentum L in the 3D case is no longer a good quantum number in 2D; instead, ν_z and M (the angular momentum projection along z) become relevant in the antisymmetrization of the two-molecule wavefunction for fermionic KRb. For example, if two KRb molecules are prepared in the same internal state and have the same ν_z, they must collide with odd M, which gives rise to a repulsive dipole–dipole interaction that suppresses the chemical reaction. On the other hand, if the two molecules have different ν_z, then the even M channel is accessible, and the reaction proceeds strongly. In the limit of a very large applied electric field and a very strong optical confinement along z, repulsive dipole–dipole interactions

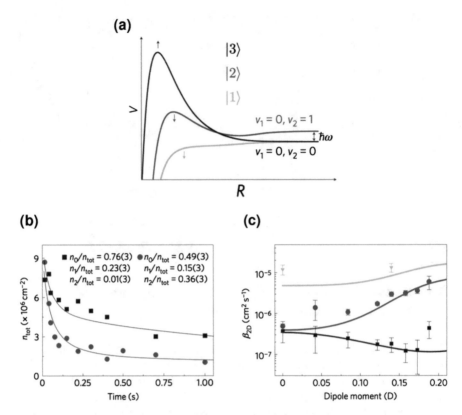

Fig. 3.6 Chemical reactions in the 1D lattice. (**a**) The barrier for collision channel 1 (green, distinguishable molecules), 2 (red, indistinguishable molecules in different bands), and 3 (black, indistinguishable molecules in the same band). Channel 3 is the most desirable to stabilize the molecular gas in a 2D geometry since it corresponds to repulsive side-by-side collisions. The arrows next to the curves indicate whether the barrier increases or decreases when the electric field increases. (**b**) Loss curves for two different initial band populations. The initial loss rate is faster when a larger fraction of molecules occupy higher bands initially. (**c**) The measured 2D loss coefficients β_{2D} for the three cases depicted in (**a**), with the same color coding. Reproduced from Ref. [7]

dominate for both Bose and Fermi quantum statistics [3, 7, 15]. In this regime, the characteristic length scale associated with the dipole–dipole interaction is larger than the harmonic oscillator length in z.

Stereo-chemical reactions in the quantum regime were explored in Ref. [7], where the reaction rate as a function of the molecular temperature, the harmonic confinement along z, and the induced dipole moment were systematically measured. Specifically, Fig. 3.6a depicts the three different configurations that were studied in the experiment [7]: (1) molecules in different internal states and same v_z, (2) spin-polarized molecules with different v_z, and (3) spin-polarized molecules with the same v_z. Collisions via channel 1 are s-wave and result in very fast loss. Channel

2 is also undesirable, since it corresponds to unsuppressed head-to-tail collisions with even M. Channel 3 is the most ideal case, as it corresponds to side-by-side collisions with odd M. Modulation of the lattice depth (parametric heating to be discussed in Chap. 4) was used to promote molecules to higher lattice bands, which sets the relative importance of channels 2 and 3. Figure 3.6b displays measured loss curves for two different initial conditions. The black curve shows the loss for a case where about 75% of the molecules were initially in the lowest band. The red curve shows the loss for a distribution where parametric heating promoted a large fraction of molecules to the second excited band. The decay curves were used to extract the loss rate coefficient β_{2D} for the 3 channels (Fig. 3.6c). As expected, β_{2D} is the smallest for channel 3. The suppression of the reaction rate via odd M, in comparison with the head-to-tail collisions, was about a factor of 60 at an induced dipole moment of 0.174 D (Fig. 3.6c).

This two-dimensional trap configuration can be ubiquitously applied to dipolar systems to control collision processes by tuning the dipole orientation relative to the collision plane. Similar experiments have recently been performed with Feshbach molecules of highly magnetic erbium atoms confined in a 1D lattice with similar goals of removing the attractive part of the dipole–dipole interaction [10]. New experiments at MIT with NaK molecules are also moving in the direction of a 1D lattice to use strong repulsion to stabilize the system against inelastic collisions (M. Zwierlein, Private communication).

3.5 Sticky Collisions: 3-Molecule Collision Channels

Given that the KRb + KRb reaction is exothermic, there is a strong interest in the scientific community in working with bialkali species that are predicted to have endothermic reactions [4, 21, 25]. However, it was predicted that there could be substantial loss in three-body collision channels [6, 9, 13, 14]. The simple intuitive picture invoked for these processes is the following: two molecules can collide and temporarily form a reaction complex prior to separating. While they are very near each other, a third molecule can come in and force the complex to a deeper bound molecular state that results in inelastic loss of all three molecules. These processes could be responsible for observed losses that are difficult to attribute to two-body loss mechanisms, but rigorous experimental confirmation of these lossy three-body processes is work in progress.

The two-body nature of the collisions discussed in the previous section is strongly supported by the observation that inelastic loss rates are ~100 times larger with non-identical fermions than with spin-polarized fermions. Let us consider the case of three-body collisions of fermionic molecules. If the fermions are spin polarized, three-body collisions will require the p-wave channel in the same way as two-body collisions. However, if the molecules are not spin polarized, the difference between two-body and three-body could become experimentally distinguishable. For example, suppose we restrict ourselves to only two unique internal states for the fermions. Then, collisions between two non-identical fermions can take place in the s-wave channel, and would not require tunneling through a barrier. However, three-body collisions will necessarily still require a p-wave channel since two of the three fermions must be identical. In this situation, therefore, we would not expect the three-body loss rate to be very different between identical and non-identical fermions. This is in stark contrast to the scenario we have presented on the KRb loss, which again points to two-body loss as the dominant loss process. The work at MIT with fermionic NaK molecules [23], which are ostensibly not reactive in the two-body channel, could shed light on three-body loss.

With bosonic molecules the situation becomes significantly more complex. Since bosons can collide in the s-wave channel, the loss rates can be extremely fast for both two-body and three-body collisions. Even bosonic molecules that are ostensibly chemically stable have been observed to decay very quickly [16, 22], and experimental signal-to-noise ratios make it very difficult to differentiate between two-body and three-body loss mechanisms. Just as for our reactive KRb molecules, these experiments can benefit from confining the molecules in zero spatial dimension achieved with a three-dimensional optical lattice to completely remove collisional losses. We thus focus on the production, stabilization, and applications of ultracold polar molecules in 3D optical lattices in the next chapter.

References

1. J. Aldegunde, H. Ran, J.M. Hutson, Manipulating ultracold polar molecules with microwave radiation: the influence of hyperfine structure. Phys. Rev. A **80**, 043410 (2009)
2. J.M. Brown, A. Carrington, *Rotational Spectroscopy of Diatomic Molecules* (Cambridge University, Cambridge, 2003)
3. H.P. Büchler, E. Demler, M. Lukin, A. Micheli, N. Prokof'ev, G. Pupillo, P. Zoller, Strongly correlated 2D quantum phases with cold polar molecules: controlling the shape of the interaction potential. Phys. Rev. Lett. **98**, 060404 (2007)
4. J.N. Byrd, J.A. Montgomery, R. Côté, Long-range forces between polar alkali-metal diatoms aligned by external electric fields. Phys. Rev. A **86**, 032711 (2012)
5. J.P. Covey, S.A. Moses, J. Ye, D.S. Jin, Controlling a quantum gas of polar molecules in an optical lattice. *Royal Society of Chemistry book chapter: "Low temperature and low energy molecular scattering"*, 2017
6. J.F.E. Croft, J.L. Bohn, Long-lived complexes and chaos in ultracold molecular collisions. Phys. Rev. A **89**, 012714 (2014)
7. M.H.G. de Miranda, A. Chotia, B. Neyenhuis, D. Wang, G. Quéméner, S. Ospelkaus, J.L. Bohn, J. Ye, D.S. Jin, Controlling the quantum stereodynamics of ultracold bimolecular reactions. Nat. Phys. **7**(6), 502–507 (2011)
8. A.R. Edmonds, *Angular Momentum in Quantum Mechanics*, 2nd edn. (Princeton University Press, Princeton, 1960)
9. A. Frisch, M. Mark, K. Aikawa, F. Ferlaino, J.L. Bohn, C. Makrides, A. Petrov, S. Kotochigova, Quantum chaos in ultracold collisions of gas-phase erbium atoms. Nature **507**, 475–479 (2014)
10. A. Frisch, M. Mark, K. Aikawa, S. Baier, R. Grimm, A. Petrov, S. Kotochigova, G. Quéméner, M. Lepers, O. Dulieu, F. Ferlaino. Ultracold dipolar molecules composed of strongly magnetic atoms. Phys. Rev. Lett. **115**, 203201 (2015)

11. K.R.A. Hazzard, B. Gadway, M. Foss-Feig, B. Yan, S.A. Moses, J.P. Covey, N.Y. Yao, M.D. Lukin, J. Ye, D.S. Jin, A.M. Rey, Many-body dynamics of dipolar molecules in an optical lattice. Phys. Rev. Lett. **113**, 195302 (2014)

12. Z. Idziaszek, P.S. Julienne, Universal rate constants for reactive collisions of ultracold molecules. Phys. Rev. Lett. **104**, 113202 (2010)

13. M. Mayle, B.P. Ruzic, J.L. Bohn, Statistical aspects of ultracold resonant scattering. Phys. Rev. A **85**, 062712 (2012)

14. M. Mayle, G. Quéméner, B.P. Ruzic, J.L. Bohn, Scattering of ultracold molecules in the highly resonant regime. Phys. Rev. A **87**, 012709 (2013)

15. A. Micheli, G. Pupillo, H.P. Büchler, P. Zoller, Cold polar molecules in two-dimensional traps: tailoring interactions with external fields for novel quantum phases. Phys. Rev. A **76**, 043604 (2007)

16. P.K. Molony, P.D. Gregory, Z. Ji, B. Lu, M.P. Köppinger, C.R. Le Sueur, C.L. Blackley, J.M. Hutson, S.L. Cornish, Creation of ultracold $^{87}Rb^{133}Cs$ molecules in the rovibrational ground state. Phys. Rev. Lett. **113**, 255301 (2014)

17. K.-K. Ni, S. Ospelkaus, M.H.G. de Miranda, A. Pe'er, B. Neyenhuis, J.J. Zirbel, S. Kotochigova, P.S. Julienne, D.S. Jin, J. Ye, A high phase-space-density gas of polar molecules. Science **322**(5899), 231–235 (2008)

18. K.-K. Ni, S. Ospelkaus, D. Wang, G. Quemener, B. Neyenhuis, M.H.G. de Miranda, J.L. Bohn, J. Ye, D.S. Jin, Dipolar collisions of polar molecules in the quantum regime. Nature **464**, 1324–1328 (2010)

19. S. Ospelkaus, K.-K. Ni, G. Quéméner, B. Neyenhuis, D. Wang, M.H.G. de Miranda, J.L. Bohn, J. Ye, D.S. Jin, Controlling the hyperfine state of rovibronic ground-state polar molecules. Phys. Rev. Lett. **104**, 030402 (2010)

20. S. Ospelkaus, K.-K. Ni, D. Wang, M.H.G. de Miranda, B. Neyenhuis, G. Quéméner, P.S. Julienne, J.L. Bohn, D.S. Jin, J. Ye, Quantum-state controlled chemical reactions of ultracold potassium-rubidium molecules. Science **327**(5967), 853–857 (2010)

21. G. Quéméner, J.L. Bohn, Dynamics of ultracold molecules in confined geometry and electric field. Phys. Rev. A **83**, 012705 (2011)

22. T. Takekoshi, L. Reichsöllner, A. Schindewolf, J.M. Hutson, C.R. Le Sueur, O. Dulieu, F. Ferlaino, R. Grimm, H.-C. Nägerl, Ultracold dense samples of dipolar rbcs molecules in the rovibrational and hyperfine ground state. Phys. Rev. Lett. **113**, 205301 (2014)

23. S.A. Will, J.W. Park, Z.Y. Yan, H. Loh, M.W. Zwierlein, Coherent microwave control of ultracold $^{23}Na^{40}K$ Molecules. Phys. Rev. Lett. **116**, 225306 (2016)

24. B. Yan, S.A. Moses, B. Gadway, J.P. Covey, K.R.A. Hazzard, A.M. Rey, D.S. Jin, J. Ye, Observation of dipolar spin-exchange interactions with lattice-confined polar molecules. Nature **501**(7468), 521–525 (2013)

25. P.S. Żuchowski, J.M. Hutson, Reactions of ultracold alkali-metal dimers. Phys. Rev. A **81**, 060703 (2010)

Chapter 4
Suppression of Chemical Reactions in a 3D Lattice

For many experiments based on the long-range interactions of polar molecules, contact between the molecules is unnecessary or even unwanted. For such experiments, a full 3D optical lattice can be used to pin molecules to individual sites where they will remain without colliding for the duration of the experiment. While research on ultracold polar molecules in optical lattices has started only recently [3, 5], promising initial experimental results have already been achieved, as will be summarized in this Chapter. I will begin with a review of optical trapping, optical lattices, and lattice techniques.

4.1 Optical Lattice Techniques

Before describing how molecules can be stabilized in a deep optical lattice, I will describe the physics of optical lattices and how they are created from optical dipole forces of off-resonant laser beams. This discussion also provides a background on optical dipole traps.

4.1.1 Optical Dipole Potentials

The oscillator model for an optical dipole potential is given by [11]

$$V_{\text{dip}} = -\frac{1}{2\epsilon_0 c} \text{Re}(\alpha) I, \qquad (4.1)$$

© Springer Nature Switzerland AG 2018
J. P. Covey, *Enhanced Optical and Electric Manipulation of a Quantum Gas of KRb Molecules*, Springer Theses, https://doi.org/10.1007/978-3-319-98107-9_4

where ϵ_0 is the electric permittivity of vacuum, c is the speed of light, α is the polarizability of the atom at the wavelength of the laser, and I is the intensity of the laser beam. At the waist of a Gaussian beam, this intensity is given by $I_0 = 2P_0/\pi \omega_{x,0}\omega_{y,0}$. The polarizability α as a function of the laser frequency ω is given by

$$\alpha(\omega) = 6\pi\epsilon_0 c^3 \frac{\Gamma/\omega_0^2}{\omega_0^2 - \omega^2 - i(\omega^3/\omega_0^2)\Gamma}, \tag{4.2}$$

where ω_0 is the resonance frequency of the atomic transition and Γ is the corresponding linewidth. Atoms have many atomic transitions, so the full polarizability is the sum over all transitions with resonance frequencies $\omega_{0,i}$. While the real part of the polarizability describes the dipole force that the light exerts on the atoms, the imaginary part describes the scattering rate of the light by the atoms, which is given by

$$\Gamma_{sc} = -\frac{1}{\hbar\epsilon_0 c}\text{Im}(\alpha)I. \tag{4.3}$$

When the laser is sufficiently detuned from any atomic transitions as is typically the case for optical dipole traps, the rotating wave approximation can be made and the dipole potential and scattering rate can be approximated by

$$V_{\text{dip}}(\mathbf{r}) \approx \frac{3\pi c^2}{2\omega_0^3}\frac{\Gamma}{\Delta}I(\mathbf{r}) \tag{4.4}$$

and

$$\Gamma_{\text{sc}}(\mathbf{r}) \approx \frac{3\pi c^2}{2\omega_0^3}\left(\frac{\Gamma}{\Delta}\right)^2 I(\mathbf{r}), \tag{4.5}$$

respectively, where $\Delta = \omega - \omega_0$ is the detuning from an atomic resonance. Again, this will be a sum over all atomic transitions. For alkali atoms in red-detuned traps (i.e., when $\Delta < 0$, such as with $\lambda = 1064$ nm) there are only two atomic transitions which need to be considered, which are nS to $nP_{1/2}$ and $nP_{3/2}$, also known as D1 and D2.

For a Gaussian trapping beam, the intensity profile $I(\mathbf{r})$ is given by

$$I(\mathbf{r}) = \frac{2P}{\pi w^2(z)}e^{-2\frac{\sqrt{x^2+y^2}}{w^2(z)}}, \tag{4.6}$$

where P is the total power, $w(z) = w_0\sqrt{1 + (z/z_R)^2}$ is the $1/e^2$ radius of the beam, and $z_R = \pi w^2/\lambda$ is the Rayleigh length for a beam propagating in the z-direction.

Fig. 4.1 An illustration of a cloud of atoms (blue) attracted to the waist of a red-detuned optical dipole trap (grey)

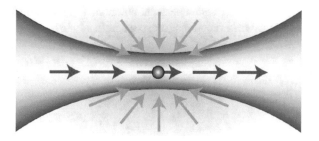

At the center of the beam (i.e., the waist), the trapping potential as illustrated in Fig. 4.1 can be approximated as [11]

$$V(r, z) \simeq -V_0 \left[1 - 2 \left(\frac{r}{\overline{w}_0} \right)^2 - \left(\frac{z}{z_R} \right)^2 \right],$$ (4.7)

where $\overline{w}_0 = \sqrt{w_{x,0} w_{y,0}}$. The trap frequencies are given by $\omega_{x_i} = \sqrt{4V_0/m w_{x_i,0}^2}$ where $x_i = x$ or y and $\omega_z = \sqrt{2V_0/m z_R^2}$.

4.1.2   Band Structure and Eigenfunctions

An optical lattice can be generated by retroreflecting a dipole trap beam upon itself to form an interference pattern, or a standing wave of nodes and antinodes. Such an interference pattern results in a periodic trapping potential, which is given by

$$V(r, z) = V_{\text{latt}} e^{-2r^2/w_{x,0} w_{y,0}} \sin(kz),$$ (4.8)

where V_{latt} is the depth of the lattice, which is maximal at an antinode for red-detuning, and $k = 2\pi/\lambda$ is the wavevector of the laser light. Near the center where $r \ll w_0$, this can be approximated by

$$V(r, z) = -V_{\text{latt}} \left(1 - 2 \frac{r^2}{w_0^2} \right) \sin^2(kz).$$ (4.9)

It is convenient to describe the lattice depth in units of recoil energies $E_r = \hbar^2 k^2/2m$. In a sufficiently deep optical lattice, the confinement on an individual lattice site can be approximated by a harmonic potential with trapping frequency

$$\omega_{\text{latt}}^2 = \frac{V_{\text{latt}}}{E_r} \frac{\hbar^2 k^4}{m^2}.$$ (4.10)

A particle in such a periodic potential $V(\mathbf{r})$ has eigenstates which can be written as Bloch wave functions (see, e.g., Refs. [1, 13])

$$\phi_q^n(x) = e^{iqx/\hbar} \cdot u_q^n(x), \tag{4.11}$$

where

$$u_q^n = \sum_l c_l^{n,q} e^{i2lkx}, \tag{4.12}$$

with l an integer. The spectrum consists of energy bands separated by forbidden energy regions (gaps). Each energy eigenstate is described by two quantum numbers: n and q. q is the quasimomentum which takes values between $-k$ and k, and n is a discrete number that labels the bands.

The energy associated with a band n can be computed using the Schrödinger equation, where the eigenstates are the Bloch wave functions u_q^n:

$$H_B u_q^n(x) = E_q^n u_q^n(x), \tag{4.13}$$

where

$$H_B = \frac{1}{2m}(\hat{p} + q)^2 + V(x). \tag{4.14}$$

While in general the lattice potential could be written as a discrete Fourier series, in the case of an optical lattice the actual potential is very well approximated by a sinusoid, and thus we can write

$$V(x) = -\frac{1}{4}V_{\text{latt}}\big(2\text{Cos}(kx) + 2\big). \tag{4.15}$$

For such a cosine potential the eigenvalues and eigenstates can be exactly written analytically in terms of Mathieu functions [24]. Numerically, the Schrödinger equation can be written in terms of the coefficients $c_l^{n,q}$

$$\sum_l H_{l,l'} c_l^{n,q} = E_q^n c_{l'}^{n,q}, \tag{4.16}$$

where $H_{l,l'} = (2l + q/\hbar k)^2 E_r$ when $l = l'$, $-1/4 \cdot V_{\text{latt}}$ when $|l - l'| = 1$, and zero otherwise [9].

The eigenstates and eigenenergies can be computed if this Hamiltonian is truncated for moderate positive and negative l. The corresponding c_l's become sufficiently small after $|l| \leq 5$ in order to accurately compute the energies of the lowest bands [9]. The eigenenergies E_q^n (in units of E_r) are shown for the first several bands in the first Brillouin zone in Fig. 4.2 for several lattice depths V_{latt} (denoted in units of E_r). The ground-band tunneling matrix element J^0 can be

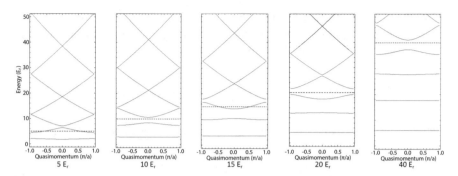

Fig. 4.2 Band structure calculations for various lattice depths. The energies of the first several bands (in units of E_r) are shown for several lattice depths as a function of the quasimomentum in the first Brillouin zone. The dashed red line shows the lattice depth at which each is calculated

calculated using the eigenenergy of the ground band E_q^0, and is represented in the tight-binding approximation (nearest-neighbor tunneling only) where the energies are of the form $-2J\cos(qa)$ [24] by

$$J^0 = \frac{1}{4}\left(\max\left(E_q^0\right) - \min\left(E_q^0\right)\right) = \frac{1}{4}\left(E_{\hbar k}^0 - E_0^0\right). \tag{4.17}$$

The tunneling matrix elements for excited bands J^n can be calculated accordingly from E_q^n in the limit of a deep lattice (a more general definition will be given below). Moreover, extending the analysis presented here to 3D is trivial in separable potentials, where the tunneling matrix elements simply become J_i^n where i denotes x, y, z.

4.1.3   Wannier Functions for Localized Atoms

With an understanding of how Bloch functions describe the motion of particles in periodic potentials of arbitrary depth V_{latt}, let us now consider specifically the case of deep lattices where particles are localized to individual lattice sites. In this limit an appropriate basis of eigenstates is given by Wannier functions, which are given by [9, 13]

$$w_n(x - x_i) = \frac{1}{\sqrt{L}}\sum_q e^{-iqx_i/\hbar}\phi_q^n(x), \tag{4.18}$$

where q is summed over the first Brillouin zone and N is the total number of lattice sites and x_i is the position of the ith lattice site in units of the lattice spacing a. The extent to which a particle is localized on a lattice site is determined by how much overlap this Wannier function has with a neighboring site. To quantify this, the rigorous way (valid even beyond the tight-binding approximation) to calculate the tunneling matrix element J^n is given by [9]

$$J^n = \int w_n(x - x_i)\left(-\frac{\hbar^2}{2m}\frac{\partial^2}{\partial x^2} + V(x)\right)w_n(x - x_j)dx \tag{4.19}$$

for neighboring sites i and j. Moreover, the on-site interaction energy U is given by

$$U^n = \frac{4\pi\hbar^2 a}{m}\int |w_n(\mathbf{r})|^4 d^3\mathbf{r}, \tag{4.20}$$

where a is the scattering length between the two particles.

When the lattice is sufficiently deep a few very useful approximations can be made. The Wannier functions can be approximated as Gaussians of RMS width $a/(\pi s^{1/4})$, where a is the lattice period, and $s = V_{\text{latt}}/E_r$ [23]. Moreover, the tunneling matrix element in the ground band can be approximated by [8]

$$J^0/E_r = 1.39666 s^{1.051} e^{-2.12104\sqrt{s}} \tag{4.21}$$

and the on-site interaction energy in the ground band can be approximated by [8]

$$U^0/E_r = 5.97(a/\lambda)s^{0.88}, \tag{4.22}$$

where λ is the wavelength of the light generating the lattice.

4.1.4   Calibrating the Lattice Depths

We use two main approaches to calibrate the depth of the lattice. The first is a parametric drive of the lattice intensity (called "parametric heating") which excites atoms from the $n = 0$ to the $n = 2$ band of the lattice. In time of flight this results in a much larger cloud. Therefore, we can study the size of the cloud as a function of the modulation frequency to learn the resonant frequency of this transition. The above discussion allows us to infer $s = V_{\text{latt}}/E_r$. This approach is very simple, but inherently less accurate because the width of the modulation resonance is broadened by the width of the two bands, and is potentially shifted to lower frequencies by modulating the lattice too hard and introducing more anharmonicity. Therefore, we typically perform the modulation at a depth of around ≈ 20 E_r for ≈ 2 ms with ≈ 5–10% modulation depth. This is the method used to determine the anisotropic polarizability of molecules in a 1D lattice, which will be described in the following section.

A more precise method is to perform Kapitza-Dirac scattering of a BEC off of the lattice potential. By flashing on the lattice potential for $\tau < 10\,\mu s$ (much shorter than any motional timescale in the lattice), the BEC is adiabatically projected onto plane-wave components (components in many bands), whose relative populations can be used to calculate the lattice depth. Specifically, the population fraction in the $\pm 2n\hbar k$ component is approximately equal to $\mathcal{J}_n^2(s E_r \tau/2\hbar)$, where $\mathcal{J}_n$ is the ordinary Bessel function of order n [7, 17]. This approximation is only valid at short times, and so we try to operate at a lattice depth sufficiently high to minimize the $n = 0$ population within a few μs. This approach gives results consistent with parametric heating, and allows us to calibrate the lattice depth to an uncertainty of $\approx 5\%$.

4.1.5   Superfluid to Mott Insulator Transition

The dynamics of ultracold bosons (such as our Rb atoms) in an external potential $V(\mathbf{x})$ is described by

$$\hat{H} = \int d^3 \mathbf{x} \hat{\psi}^\dagger(\mathbf{x})\left(-\frac{\hbar^2}{2m}\nabla^2 + V(\mathbf{x})\right)\hat{\psi}(\mathbf{x}) + \frac{1}{2}\frac{4\pi a\hbar^2}{m}$$

$$\int d^3\mathbf{x}\hat{\psi}^\dagger(\mathbf{x})\hat{\psi}^\dagger(\mathbf{x})\hat{\psi}(\mathbf{x})\hat{\psi}(\mathbf{x}) - \mu\int d^3\mathbf{x}\hat{\psi}^\dagger(\mathbf{x})\hat{\psi}(\mathbf{x}), \qquad (4.23)$$

where $\hat{\psi}$ and $\hat{\psi}^\dagger$ are bosonic field operators. In the case of a lattice potential they can be expanded into the Wannier functions discussed above, as

$$\hat{\psi}(\mathbf{x}) = \sum_i \hat{a}_i w(\mathbf{x} - \mathbf{x}_i), \qquad (4.24)$$

where $\hat{a}_i$ ($\hat{a}_i^\dagger$) are bosonic annihilation (creation) operators on the site i. These operators obey the canonical commutation relation $[\hat{a}_i, \hat{a}_j^\dagger] = \delta_{ij}$.

This expansion allows us to rewrite the Hamiltonian in terms of the parameters discussed above, J and U, and in the tight-binding limit the Hamiltonian simplifies to the so-called Bose-Hubbard Hamiltonian

$$\hat{H} = -J\sum_{<i,j>}\hat{a}_i^\dagger\hat{a}_j + \sum_i(\epsilon_i - \mu)\hat{n}_i + \sum_i\frac{1}{2}U\hat{n}_i(\hat{n}_i - 1), \qquad (4.25)$$

where $\hat{n}_i = \hat{a}_i^\dagger\hat{a}_i$ is the number operator that counts the number of bosons on the lattice site i, $\epsilon_i = V_{\text{ext}}(x_i)$ is a site-by-site energy shift, for example, coming from an additional harmonic confinement generated by the lattice beams and dipole trap beams V_{ext}, and μ is the chemical potential of the atoms in the trap. If V_{ext}

is too large compared to U, the system will prefer to put more atoms on the sites in the middle rather than putting them further out where ϵ_i is larger. The third term describes this energy cost for multiple occupancy, as indicated by the presence of U.

This Hamiltonian exhibits a quantum phase transition from a superfluid to a Mott insulator which was first observed in cold atoms in 2002 [10]. The evidence of a superfluid is phase coherence between all the sites, which gives rise to an interference pattern upon a sudden release of the atoms from the trap (see Fig. 4.3a). Once the Mott insulating phase is reached, the phase coherence is lost, and there is no longer interference upon expansion. In fact, a superfluid can be ramped into the Mott insulator phase and then back into the superfluid phase, and the phase coherence returns. These tools will be discussed in Chap. 6 where we describe how we used them to determine if the Rb Mott insulator is behaving properly. The phase transition should occur for Rb in a lattice of wavelength $\lambda = 1064$ nm at ~ 16 E_r, which provides another way to diagnose the system, particularly in the presence of K even for zero K-Rb interaction [2]. This is shown in Fig. 4.3b, d by plotting the visibility (i.e., the number of atoms at the position of a diffraction peak compared with the number of atoms half way between two adjacent diffraction peaks) of the interference peaks versus the lattice depth.

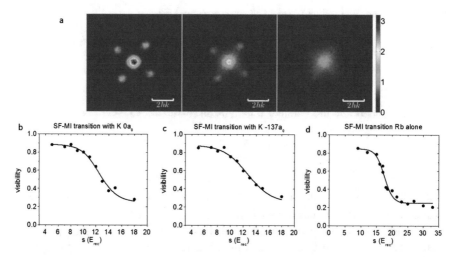

Fig. 4.3 The superfluid to Mott insulator transition of Rb with K. (**a**) shows the disappearance of interference beams in crossing the superfluid to Mott insulator phase transition for Rb alone. The three images are 12, 17, and 22 E_r, from left to right. Reproduced from Ref. [18]. (**b**) and (**c**) show the visibility of the interference peaks as a function of lattice depth in the presence of K for $a_{\text{K-Rb}} = 0$, and -137 a_0, respectively. (**d**) shows the visibility in the case of Rb alone, which corresponds to images as in (**a**)

4.1.6 Band Mapping

For the experiments described in this thesis, the goal was to load the atoms and molecules into the ground band of the optical lattice. However, due to the finite temperature of the atoms and nonadiabaticity during the loading, we often have atoms or molecules in the first excited band. Identifying the higher band fraction is particularly important for fermions, such as ^{40}K and KRb, where double occupancy in a spin-polarized sample can only occur through higher band excitations. The standard way to detect the fraction of atoms in higher bands is called band mapping, by which the different bands are mapped onto different positions upon expansion. Band mapping works by mapping each band into the quasimomentum corresponding to a Brillouin zone. To do this, it is important that the lattice intensity is ramped off quickly compared to the motion of a lattice site yet still slowly enough to be adiabatic with respect to the band gap Δ. Band mapping is an incredibly useful tool, and has been described in a number places, such Refs. [6, 9, 17, 20].

We found the refocusing technique described in Ref. [19] to be particularly useful and provided the sharpest images which were easy to analyze. In this approach, the lattice intensity is ramped to zero as discussed, but the underlying harmonic confinement (optical dipole traps) is left on. After waiting for a quarter period in this harmonic potential, the initial momentum distribution is perfectly mapped onto the position distribution. The resulting size of the first Brillouin zone in position space is given by $2\hbar k/(m\omega_{hc})$, where ω_{hc} is the oscillation frequency in the harmonic confinement potential. The ground band is from $x = -1\hbar k/(m\omega_{hc})$ to $1\hbar k/(m\omega_{hc})$, and the first excited band is from $x = -2\hbar k/(m\omega_{hc})$ to $-1\hbar k/(m\omega_{hc})$ and $1\hbar k/(m\omega_{hc})$ to $2\hbar k/(m\omega_{hc})$. This analysis can be applied to two dimensions, as in Fig. 4.4, and can thus be generalized to estimate the full 3D ground band fraction. I will come back to this topic in Chap. 6.

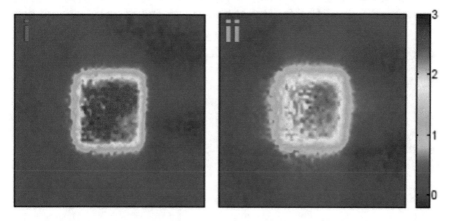

Fig. 4.4 Band mapping of K in the lattice. The two cases show a lower excited band fraction in **i** and a higher excited band fraction in **ii**. Reproduced from Ref. [4]

4.2 Long-Lived Molecules in a 3D Optical Lattice

I will now discuss molecules in an optical lattice, which can be obtained by either
loading a bulk gas of molecules into the lattice or by loading K and Rb into the
lattice and then making molecules on sites that happen to have one (and only one)
of each. Both approaches tend to yield $3\text{--}4 \times 10^4$ molecules in the lattice with a
filling fraction of ~5%. This is because given our trapping conditions the entropy
per particle in a Fermi gas of molecules at $T = T_F$ roughly matches 5% filling
in the lattice, and a Bose-Fermi mixture of atoms at $T = T_c, T_F$ loaded into the
lattice tend to have ~5% of sites with one of each in our specific case (this is not a
generic argument; depends on the number of atoms, loading conditions, and external
trapping conditions, etc.). Therefore, both approaches were used in the early years
of our work with optical lattices. However, by the next chapter we were exclusively
using the latter approach. The quantum synthesis idea described in Chap. 6 is in fact
based entirely on this latter idea, and the ability to put one K and Rb in every site
using atomic insulators.

Two years before the first production of ground-state polar molecules,
researchers had already prepared homonuclear Feshbach molecules in an optical
lattice by magneto-association of pairs of atoms contained in single sites of a
bosonic Mott insulator of Rb [26]. They were then able to remove unpaired atoms,
leaving only sites that were either empty or contained a Feshbach molecule. In
such an experiment, no collisions between molecules were allowed in a deep
lattice and therefore molecules could not undergo inelastic scattering via vibrational
quenching from the Feshbach state. The lifetime was measured to be 700 ms, which
is significantly longer than that in a bulk gas where molecules can collide with atoms
and with each other [26]. However, there is no fundamental limit to this lifetime,
and it was ultimately limited by inelastic light scattering from the trapping light.

More recently, we have demonstrated the creation of KRb Feshbach and ground-
state molecules in 3D optical lattices [3], and have shown that their lifetimes
can exceed 20 s (see Fig. 4.5). This lifetime is also limited by off-resonant light

Fig. 4.5 Long lifetimes are
experienced, after a fast
initial loss attributed to dark
molecules and residual
unpaired atoms. The inset
shows that this long lifetime
is independent of the electric
field. Reproduced from
Ref. [3]

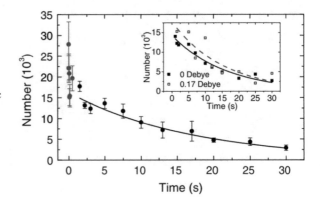

scattering from the optical trap, which is fairly far detuned from molecular resonances. The lifetime of these molecules in the optical lattice was shown to be significantly longer than the timescales that are relevant for the dominant interaction energies between the molecules [12, 15], and so this offers an excellent starting point for the investigation of dipole–dipole interactions in an optical lattice and for the engineering of a lattice model with a spin-1/2 Hamiltonian.

As discussed above, applying an electric field in the laboratory frame can potentially reduce the lifetime of the molecular gas by allowing the attractive dipolar interaction to reduce the p-wave barrier. In the case of a deep 3D lattice and therefore the absence of tunneling, the short-range collisional physics is fully suppressed, and hence the mechanism for reducing the lifetime by an electric field is no longer valid. Indeed, we observed experimentally that the lifetime of the molecules in a deep 3D lattice remains the same with an electric field of $\sim$4 kV/cm as with no field (see Fig. 4.5). This was very encouraging; we can now turn on the full-strength dipole–dipole interactions with electric fields without having to worry about any extra loss our molecular sample may experience [3].

4.3   The Continuous Quantum Zeno Mechanism: Stability from Strong Dissipation

I now consider molecules in a shallow lattice in which there is significant tunneling. The largest measured two-body loss rate for a bulk gas of KRb at zero electric field was for the rotational mixture discussed in Chap. 3. In this case, s-wave collisions are allowed and so the on-site loss rate in a 3D lattice would be very large, even larger than both the tunneling rate and the energy gap between the two lowest lattice bands. In this regime, however, the process for a molecule to tunnel onto a lattice site that is already occupied by another molecule is greatly suppressed by the continuous quantum Zeno effect [25], where the effective loss rate of the whole system counterintuitively decreases as the on-site loss rate increases. In other words, as the loss rate of two distinguishable molecules on a single lattice site increases, tunneling onto an occupied site is suppressed, as if the molecules knew they would undergo inelastic loss on that site.

The experimental setup for the work reported in Refs. [27, 28] is shown schematically in Fig. 4.6. About 10^4 ground-state KRb molecules were produced in a deep 3D lattice (along all three directions). An incoherent mixture of $\mid \downarrow \rangle = \mid 0, 0, -4, 1/2 \rangle$ and $\mid \uparrow \rangle = \mid 1, -1, -4, 1/2 \rangle$ ($\mid N, m_N, m_K, m_{Rb} \rangle$) was created by first applying a $\pi/2$ pulse between these two rotational states and then waiting 50 ms for the superposition to decohere due to stray fields and the trapping light. The two rotational states constitute a pseudo-spin-1/2 system, where the Hilbert space consists of two orthogonal states (i.e., described by Pauli matrices). Then, the lattice depth along y was reduced within 1 ms to allow tunneling with a rate J_y/h given by the tunneling matrix element between neighboring sites J. This created

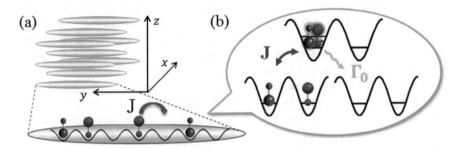

Fig. 4.6 (a) Experimental setup for studying the continuous quantum Zeno effect. The molecules are held in a 3D lattice, with strong lattices along x and z and a weak lattice along y. This realizes a system of decoupled 1D tubes with a weak corrugation along the tubes. (b) The tunneling energy along the tube direction is J and the on-site loss rate is Γ_0. Reproduced from Ref. [28]

a system of decoupled 1D tubes with a weak lattice along the tube axis (y). After holding the molecules in the lattice for a variable amount of time, the number of molecules in the $N = 0$ state ($| \downarrow \rangle$) was measured to determine the loss of this mixed-spin system.

There is no Pauli suppression preventing a $| \downarrow \rangle$ and a $| \uparrow \rangle$ molecule from occupying the same lattice site, and their on-site loss rate Γ_0 is very large. In fact, $\Gamma_0 \gg \frac{J_y}{\hbar}$. In this limit, tunneling-induced on-site loss becomes a second-order perturbative process and the effective loss rate for the overall system becomes

$$\Gamma_{\text{eff}} = \frac{2J_y^2}{\hbar^2 \Gamma_0}. \tag{4.26}$$

Since particles are lost via a two-body process [28],

$$\dot{N}_i(t) = -\frac{\kappa N_i(t)^2}{N_i(0)}, \tag{4.27}$$

where $i = | \uparrow \rangle$ or $| \downarrow \rangle$, $\kappa = 4q\Gamma_{\text{eff}}n_\downarrow(0)$ is the loss rate coefficient, and $N_i(0)$ is the initial number of molecules in state i. $n_\downarrow(0)$ is the initial filling fraction of $| \downarrow \rangle$ molecules, so $2n_\downarrow(0)$ is the total filling fraction, and q is the number of nearest neighbors. Measuring the dependence of κ on J_y and Γ_0 is thus directly achievable in the experiment. An extra benefit is the determination of the molecule filling fraction in the 3D lattice.

The transverse lattice depth $V_\perp$ was typically $40\,E_R$, while the lattice along y direction was varied between 5 and 16 E_R. Here, $E_R = \hbar^2 k^2/2m$ is the lattice photon recoil energy for KRb. To vary Γ_0 only, the transverse lattice depths were changed. To vary J_y, the lattice depth along the y direction was changed; however, to keep Γ_0 fixed, the transverse lattice depths were also changed accordingly.

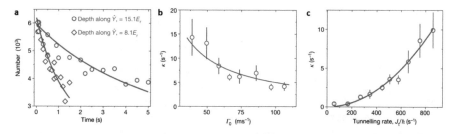

Fig. 4.7 (**a**) Typical loss curves for two different lattice depths along the y-direction, where the data is fit to the solution of the rate equation (Eq. (4.27)). (**b**) and (**c**) The dependence of κ on Γ_0 (**b**) and $J_y \hbar$ (**c**). Reproduced from Ref. [27]

Figure 4.7a shows a typical loss curve for $V_y = 5E_R$ and $V_\perp = 25E_R$. The data was fit to the solution of Eq. (4.27). Naïvely calculating the filling using Eq. (4.27) provided an overestimate of the filling fraction arising from the modification of the single-particle Wannier orbitals due to interactions which are not taken into account in this equation. Since Γ_0 is larger than or on the same order of magnitude as the band gap, higher band populations must be included to reach the correct value of Γ_0. For the experimental parameters, properly accounting for higher bands can decrease Γ_0 by about a factor of 5, which then increases Γ_{eff} by the same factor. The curve in Fig. 4.7b clearly demonstrates that the loss rate coefficient κ and also the effective loss rate Γ_{eff} decrease as Γ_0 increases, while the data in Fig. 4.7c shows that κ and Γ_{eff} have roughly a quadratic dependence on $J_y/\hbar$. These scalings are in accordance with Eq. (4.26). These results provide strong evidence for the continuous quantum Zeno effect, and at the same time allowed us to determine that the lattice filling fraction was around 6%.

4.4 Anisotropic Polarizability

The long lifetime for molecules in a deep lattice allows for investigations of the real part of the molecular AC polarizability in addition to its lossy imaginary component. This is important for us to understand the molecular energy level shifts inside the lattice for the implementation of the lattice spin model (see Chap. 5). The wavefunctions for many molecular states are spatially anisotropic, and hence the corresponding polarizability depends on the angle between the polarization of the lattice light and the quantization field defined by an applied electric or magnetic field (see Fig. 4.8). By tuning this angle carefully, KRb molecules are found to exhibit a "magic" trapping condition in which the polarizability of two rotational states (as used to define a spin-1/2 system) can be adjusted to match with each other [14, 21] (see Fig. 4.9).

Fig. 4.8 (**a**) The geometry of the experiment with a 1D lattice and the quantization field set by the magnetic field. (**b**) The energy level diagram of the relevant rotational states. (**c**) The effect of a differential polarizability on the rotational state transition energy. Reproduced from Ref. [21]

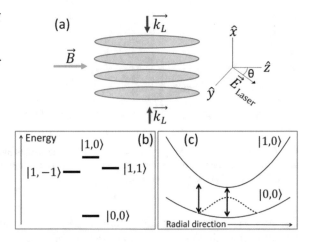

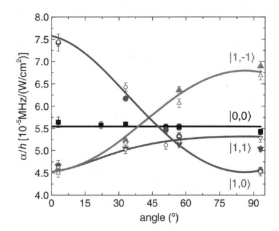

Fig. 4.9 The ac polarizability of KRb at 1064 nm for the $|0, 0\rangle$ (black squares), $|1, 0\rangle$ (blue circles), $|1, 1\rangle$ (red inverted triangles), and $|1, -1\rangle$ (green triangles) states ($|N, m_N\rangle$) as a function of the angle between the quantization field (B-field) and the polarization of the light field in a 1D optical lattice. Reproduced from Ref. [21]

Such "magic" conditions are reached in optical lattice clocks by adjusting the lattice light wavelength and are routinely used to minimize the effects of the trapping light on the clock transition [16]. For molecules in this "magic" angle lattice, the coherence time for a coherent superposition of the two rotational states is found to be orders of magnitude longer than otherwise (see Fig. 4.10), which is an important prerequisite for studying spin physics with polar molecules in an optical lattice. Note that a coherence time of roughly 1 s has been observed recently between two hyperfine states of NaK in a single rotational state in an optical dipole trap [22].

Fig. 4.10 The Ramsey coherence time is sharply peaked at the magic angle, and the inset shows the Ramsey fringe as a function of dark time. Reproduced from Ref. [21]

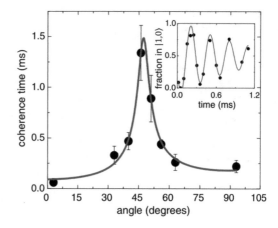

4.5 Hyperpolarizability

In addition to an anisotropic polarizability, the molecules suffer from significant hyperpolarizability, which is a nonlinear optical property originating from a second-order electric susceptibility of polar molecules. This second-order polarizability is what gives rise to second-harmonic generation in molecular samples. Essentially, the polarizability itself is a function of the beam intensity such that the trap depth V_{dip} is given by

$$V_{\mathrm{dip}}(I) = -\frac{1}{2\epsilon_0 c}\alpha(I)I, \tag{4.28}$$

where $\alpha(I) = \alpha_0 + \alpha_1 I$ with α_0 as described in the previous section.

This hyperpolarizability is significant because the trap has an intensity curvature across the cloud. Therefore, even if the differential polarizability is zero at a certain lattice depth, the variation of the intensity across the cloud induces a finite differential shift. As shown in Fig. 4.8c, a non-magic trap leads to a differential trapping potential between the two states of the form $\Delta V = \Delta \alpha I$. Conversely, the hyperpolarizability gives rise to a term that looks like $\Delta V = \Delta \alpha(I)I$ even if the trap is magic at a given intensity. Since the $\alpha(I)$ is different between the two states, there is a net polarizability difference across the trap. In practice we find that the total ac Stark shift across the entire cloud can be limited to only ≈ 1 kHz, which is $\ll 10$ Hz per site at the center.

Figure 4.11 shows the differential ac Stark shift as a function of the lattice depth in units of E_r. Their nonlinearity is evidence of the hyperpolarizability of the states involved. The inset shows the differential polarizability of the $|0, 0\rangle$ to $|1, -1\rangle$ transition between 20 to 60 E_r, and is zero when the lattice depth is about 40 E_r in each beam. However, the differential light shift has zero slope at 20 E_r, so the effect

Fig. 4.11 The differential
light shift (with respect to the
$|0, 0\rangle$ state is shown as a
function of the lattice depth.
Green, black, and blue lines
show calculated light shifts
for the $|1, 0\rangle$, $|1, 1\rangle$, and
$|1, -1\rangle$ rotational states,
respectively. Red points are
experimental data for the
$|1, -1\rangle$ state. Inset: An
expanded view for the $|1, -1\rangle$
state is displayed.
Reproduced from Ref. [27]

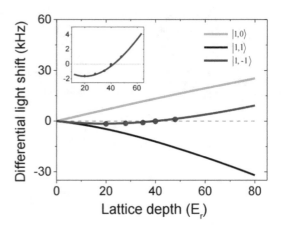

of the intensity curvature across the cloud can be mitigated at this lattice depth.
Note that the differential light shift of the transition to $|1, 0\rangle$ never has a vanishing
derivative.

4.6   Polarizabilities at Other Wavelengths

We have long been interested in using other wavelengths for trapping the molecules.
There are two primary reasons for this: (1) since the dipolar interaction scales as
$1/R^3$, reducing the lattice spacing by a factor of 2 increases the interaction energy
by a factor of 8! (2) it may be easier to make a colder, denser, lower-entropy gas
of molecules when starting from atomic gases trapped by wavelengths other than
$\lambda = 1064$ nm. I will now discuss our efforts to use other wavelengths, and our
measurements of the atomic and molecular polarizabilities at these wavelengths.

4.6.1 $\lambda_2 = 532$ nm

A factor of eight increase in the interaction energy of the molecules in the lattice
would put us at $J_\perp/2 \sim 1$ kHz (see Chap. 5), where we could even start to enter
the blockade regime of excitations with Rabi frequencies ~ 1 kHz or less. Such
interaction energies could make spin diffusion and impurity dynamics much easier
to observe. Moreover, realizing itinerant models such as the t-J model based on the
stability afforded by the continuous quantum Zeno mechanism would be much more
straightforward. Further, high power lasers are readily available at this wavelength,
and we used a Verdi by Coherent.

Additionally, since $\lambda_2 = 532$ nm is blue-detuned for both K and Rb, a $\lambda_2 = 532$ nm could supplement the $\lambda_1 = 1064$ beams already in use, and serve to

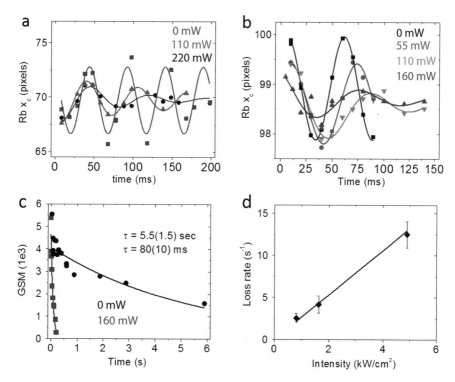

Fig. 4.12 Reducing the harmonic confinement using $\lambda_2 = 532$ nm dipole traps. (**a**) Increasing the power of a $\lambda_2 = 532$ nm dipole aligned onto the $\lambda_1 = 1064$ nm dipole trap reduces the combined trap frequency. (**b**) The addition of the λ_2 beam increases the anharmonicity, and the description of the trapping potential becomes more complicated. (**c**) The lifetime of ground state molecules in the λ_1 lattice, with and without the λ_2 dipole beam on, whose $1/e^2$ radius is $\sim 50\,\mu$m. The lifetime with it is $\sim 100\times$ shorter. (**d**) The loss rate of molecules in the lattice with the λ_2 beam as a function of its intensity

decrease the harmonic confinement since λ_2 is anti-confining. The significance of this will become more clear in Chap. 6, but a flatter trap with less harmonic confinement would allow a larger region in the lattice where molecules could be created efficiently. This effect is shown in Fig. 4.12a, where increasing the λ_2 beam power decreases the trap frequency of the combined trap. However, the combined trap is now a more complicated potential which is very sensitive to alignment. Figure 4.12b highlights that the addition of the λ_2 beam substantially increases the damping rate, and depends strongly on the power. The combined trap is very anharmonic and its properties are prone to enormous fluctuation. Therefore, while this could potentially be useful, we ultimately found it to be too difficult to use as a part of the lattice loading.

Perhaps even more important is the lifetime of the molecules in a λ_2 trap. To measure this, we loaded ground state molecules into the λ_1 lattice and turned on a λ_2 beam. The results are shown in Fig. 4.12c and d. The λ_2 $1/e^2$ beam radius is $\approx 50\,\mu$m, and it reduces the lifetime by a factor of almost 100. The loss rate is proportional with the λ_2 beam intensity, suggesting that is entirely from $\text{Im}(\alpha_{\text{KRb}}(532))$ being very large. These observations corroborate calculations by Svetlana Kotochigova, and the loss rate is prohibitively high for most experiments we would like to do. Thus, we never pursued 532 nm wavelength dipole traps or lattices further for either anticonfinement or an enhanced dipolar interaction strength.

4.6.2 $\lambda_3 = 755\,nm$

Another intriguing wavelength on the blue-detuned side is $\lambda_3 = 755$ nm, where K and Rb would feel the same lattice depth in units of their respective E_r. Since this lattice has a smaller lattice constant, each lattice site is smaller, which increases the on-site interaction energy U, discussed above. Therefore, such a lattice could accommodate more atoms in the $n = 1$ Mott shell before the onset of double occupancy at $\mu/V > 1$. Moreover, the anti-harmonic confinement of this beam could (at least partially) cancel the harmonic confinement of the $\lambda_1 = 1064$ nm dipole traps, thereby further increasing the number of atoms in the $n = 1$ shell. The significance of the large single occupancy Mott shell will become more clear in Chap. 6.

We studied this wavelength using a tunable Eagleyard ECDL and TA system operating between ~ 750–770 nm. Figure 4.13a and b shows the slosh of K and KRb clouds, respectively, in a λ_1 dipole trap after a ≈ 1 ms pulse of the λ_3 beam. Note that the amplitude is $\approx 20\times$ smaller for KRb than it is for K, but also the sign is opposite. Therefore, while λ_3 is blue detuned for K and Rb, it is red-detuned for KRb. Figure 4.13c shows the same measurement for K, Rb, and KRb all on the same plot. The ratio of the K and Rb amplitudes matches the difference in their polarizabilities at λ_3 and their mass ratio.

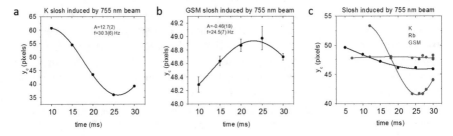

Fig. 4.13 The polarizabilities at $\lambda_3 = 755$ nm. (**a**) The slosh of K in the $\lambda_1 = 1064$ trap from a ~ 1 ms pulse of the $\lambda_3 = 755$ nm trap. (**b**) The slosh of KRb from the same pulse from the λ_3 beam. Note that the slosh of KRb has the opposite sign as the slosh of K. (**c**) The slosh of K, Rb, and KRb all on the same plot induced from the λ_3 pulse

The scenario at λ_3 is quite the opposite as λ_1 where the polarizability is much larger and the same sign for KRb as K and Rb. At λ_3 the polarizability is opposite but also much smaller for KRb. Moreover, this situation is nearly the same at 759 and 763 nm. Further, the lifetime of KRb in the λ_1 and λ_3 trap is similar to what it is with just λ_1, which is ≈ 1 s. These observations suggest that the polarizability for KRb may be shallowly crossing zero, and thus be blue-detuned on the blue of λ_3. Unfortunately, we were unable to confirm this. However, it is unclear how useful this trap would be since the lattice depth would be very low at realistic beam intensities.

A large window of low $\mathrm{Re}(\alpha)$ and $\mathrm{Im}(\alpha)$ is quite uncommon for molecules, which have dense spectra at most wavelengths. We have compared our results to polarizability calculations by Svetlana Kotochigova, which demonstrate a similar behavior. Nevertheless, a qualitative comparison is lacking to date, and a thorough understanding of this issue has not been reached. Yet, enough is known that we may decide to investigate this further later on to take advantage of the long molecular lifetime in optical traps at this wavelength.

4.6.3 $\lambda_4 = 790\,nm$

Wavelengths near the atomic resonances of K and Rb allow species selective manipulation, and $\lambda_4 = 790$ provides a red-detuned trap for just K where the Rb polarizability crosses zero. This tool could be very useful for tailoring the atomic gases to efficiently make molecules, and I will return to this topic at the beginning of Chap. 6. Because of the different mass, trapping frequency, and quantum statistics of K and Rb, they tend to have different positions, sizes, and momentum distributions at cold temperatures in weak traps. Therefore a trap that only manipulates K could be very useful for enhancing the number of molecules that could be produced in the quantum degenerate regime.

To study the effects of this trap we used a tunable Eagleyard ECDL and TA system operating between $\sim$785–795 nm. Figure 4.14 shows the position of the K and Rb clouds as the $\lambda_4 = 790$ nm beam is scanned across the $\lambda_1 = 1064$ nm beam. Since the trap is species-selective, Rb has very little change while K moves substantially. The dispersive effect on the position of K can be seen in Fig. 4.14b as the λ_4 beam is scanned for two different intensities. This capability allows us to cancel the differential sag, and by increasing the power we can squeeze the large K Fermi gas so that is better spatially matched with the Rb BEC.

Although the absolute laser stability is not very important, we locked the lasers to a HighFinesse wavemeter, and carefully ensured single-mode laser operation. Figure 4.14c shows the lifetime of Feshbach molecules in a λ_4 dipole trap for ≈ 60 mW with a $1/e^2$ beam radius of $\approx 100\,\mu$m. The lifetime is incredibly short, ≈ 1 ms, which is perhaps not surprising given the proximity of the wavelength to the enormous number of rovibronic excited states. The same measurement is carried

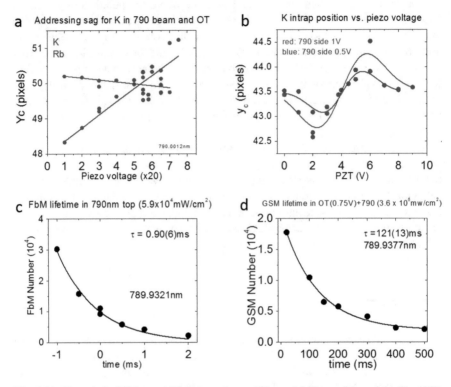

Fig. 4.14 The polarizabilities and lifetimes at $\lambda_4 = 790$ nm. (**a**) The position of the K and Rb clouds as the $\lambda_4 = 790$ nm beam is scanned across the $\lambda_1 = 1064$ nm beam. Since the trap is species-selective, Rb has very little change while K moves substantially. (**b**) The dispersive effect on the position of K as the λ_4 beam is scanned for two different intensities. (**c**) The lifetime of Feshbach molecules in the combined λ_1/λ_4 trap. This lifetime is prohibitively short. (**d**) The same measurement, but for ground state molecules. This lifetime is $\sim 100\times$ longer, but still too short for most experiments

out with ground state molecules in Fig. 4.14d, and the lifetime is $\approx 100\times$ longer. As discussed in Chap. 6, such short lifetimes for the Feshbach molecules render this wavelength effectively unusable.

References

1. N.W. Ashcroft, N.D. Mermin, *Solid State Physics* (Saunders College Publishing, Fort Worth, 1976)
2. Th. Best, S. Will, U. Schneider, L. Hackermüller, D. van Oosten, I. Bloch, D.-S. Lühmann, Role of interactions in ^{87}Rb–^{40}K bose-fermi mixtures in a 3d optical lattice. Phys. Rev. Lett. **102**, 030408 (2009)

3. A. Chotia, B. Neyenhuis, S.A. Moses, B. Yan, J.P. Covey, M. Foss-Feig, A.M. Rey, D.S. Jin, J. Ye, Long-lived dipolar molecules and feshbach molecules in a 3D optical lattice. Phys. Rev. Lett. **108**, 080405 (2012)

4. J.P. Covey, S.A. Moses, M. Garttner, A. Safavi-Naini, M.T. Miecnkowski, Z. Fu, J. Schachenmayer, P.S. Julienne, A.M. Rey, D.S. Jin, J. Ye, Doublon dynamics and polar molecule production in an optical lattice. Nat. Commun. **7**, 11279 (2016)

5. J.G. Danzl, M.J. Mark, E. Haller, M. Gustavsson, R. Hart, J. Aldegunde, J.M. Hutson, H.-C. Nägerl, An ultracold high-density sample of rovibronic ground-state molecules in an optical lattice. Nat. Phys. **6**, 265–270 (2010)

6. M.H.G. de Miranda, Control of dipolar collisions in the quantum regime. PhD thesis, University of Colorado, Boulder, 2010

7. B.R. Gadway, Bose gases in tailored optical and atomic lattices. PhD thesis, Stony Brook University, 2012

8. F. Gerbier, A. Widera, S. Fölling, O. Mandel, T. Gericke, I. Bloch, Interference pattern and visibility of a mott insulator. Phys. Rev. A **72**, 053606 (2005)

9. M. Greiner, Ultracold quantum gases in three-dimensional optical lattice potentials. PhD thesis, Ludwig-Maximilians-Universität München, 2003

10. M. Greiner, O. Mandel, T. Esslinger, T.W. Hansch, I. Bloch. Quantum phase transition from a superfluid to a Mott insulator in a gas of ultracold atoms. Nature **415**(6867), 39–44 (2002)

11. R. Grimm, M. Weidemller, Y.B. Ovchinnikov, Optical dipole traps for neutral atoms. Adv. Atom. Mol. Opt. Phys. **42**, 95–170 (2000)

12. K.R.A. Hazzard, S.R. Manmana, M. Foss-Feig, A.M. Rey, far-from-equilibrium quantum magnetism with ultracold polar molecules. Phys. Rev. Lett. **110**, 075301 (2013)

13. C. Kittel, *Quantum Theory of Solids* (Wiley, New York, 1963)

14. S. Kotochigova, D. DeMille, Electric-field-dependent dynamic polarizability and state-insensitive conditions for optical trapping of diatomic polar molecules. Phys. Rev. A **82**, 063421 (2010)

15. M. Lemeshko, R.V. Krems, H. Weimer, Nonadiabatic preparation of spin crystals with ultracold polar molecules. Phys. Rev. Lett. **109**, 035301 (2012)

16. A.D. Ludlow, M.M. Boyd, J. Ye, E. Peik, P.O. Schmidt, Optical atomic clocks. Rev. Mod. Phys. **87**, 637–701 (2015)

17. S.A. Moses, A quantum gas of polar molecules in an optical lattice. PhD thesis, University of Colorado, Boulder, 2016

18. S.A. Moses, J.P. Covey, M.T. Miecnikowski, B. Yan, B. Gadway, J. Ye, D.S. Jin, Creation of a low-entropy quantum gas of polar molecules in an optical lattice. Science **350**(6261), 659–662 (2015)

19. P.A. Murthy, D. Kedar, T. Lompe, M. Neidig, M.G. Ries, A.N. Wenz, G. Zürn, S. Jochim. Matter-wave fourier optics with a strongly interacting two-dimensional fermi gas. Phys. Rev. A **90**, 043611 (2014)

20. B. Neyenhuis, Ultracold polar krb molecules in optical lattices. PhD thesis, University of Colorado, Boulder, 2012

21. B. Neyenhuis, B. Yan, S.A. Moses, J.P. Covey, A. Chotia, A. Petrov, S. Kotochigova, J. Ye, D.S. Jin, Anisotropic polarizability of ultracold polar $^{40}K^{87}Rb$ molecules. Phys. Rev. Lett. **109**, 230403 (2012)

22. J.W. Park, Z.Y. Yan, H. Loh, S.A. Will, M.W. Zwierlein, Second-scale nuclear spin coherence time of trapped ultracold $^{23}Na^{40}K$ (2016). Arxiv:1606.04184v1

23. P. Pedri, L. Pitaevskii, S. Stringari, C. Fort, S. Burger, F.S. Cataliotti, P. Maddaloni, F. Minardi, M. Inguscio, Expansion of a coherent array of bose-einstein condensates. Phys. Rev. Lett. **87**, 220401 (2001)

24. C.A. Regal, Ultracold bosonic atoms in optical lattices. PhD thesis, University of Maryland, College Park, 2004

25. N. Syassen, D.M. Bauer, M. Lettner, T. Volz, D. Dietze, J.J. García-Ripoll, J.I. Cirac, G. Rempe, S. Dürr, Strong dissipation inhibits losses and induces correlations in cold molecular gases. Science **320**(5881), 1329–1331 (2008)

26. G. Thalhammer, K. Winkler, F. Lang, S. Schmid, R. Grimm, J. Hecker Denschlag, Long-lived feshbach molecules in a three-dimensional optical lattice. Phys. Rev. Lett. **96**, 050402 (2006)
27. B. Yan, S.A. Moses, B. Gadway, J.P. Covey, K.R.A. Hazzard, A.M. Rey, D.S. Jin, J. Ye, Observation of dipolar spin-exchange interactions with lattice-confined polar molecules. Nature **501**(7468), 521–525 (2013)
28. B. Zhu, B. Gadway, M. Foss-Feig, J. Schachenmayer, M.L. Wall, K.R.A. Hazzard, B. Yan, S.A. Moses, J.P. Covey, D.S. Jin, J. Ye, M. Holland, A.M. Rey, Suppressing the loss of ultracold molecules via the continuous quantum zeno effect. Phys. Rev. Lett. **112**, 070404 (2014)

Chapter 5
Quantum Magnetism with Polar Molecules in a 3D Optical Lattice

With chemical reactions under control and the molecular gas stabilized in the 3D optical lattice, we have recently begun to study a conservative manifestation of the dipole–dipole interaction, where the molecules never come in contact with each other. This is a new experimental regime where the internal degrees of freedom (spins) of molecules interact strongly, but the motional degrees of freedom are largely decoupled from the system dynamics. Certainly this is not the case for cold, ground-state alkali atoms in a lattice whose interactions are primarily short-range, although recently experiments with highly magnetic atoms and Rydberg atoms have demonstrated similar long-range interactions [1, 3].

5.1 The Dipolar spin-1/2 XXZ Hamiltonian

References [2, 5] proposed that polar molecules can be used to study quantum magnetism, and dipolar interactions should be observable even at low lattice fillings and high entropies [6]. Specifically, by defining a spin-1/2 degree of freedom using the lowest two rotational states ($| \downarrow \rangle \equiv |0, 0\rangle$ and $| \uparrow \rangle \equiv |1, -1\rangle$ or $| \uparrow \rangle \equiv |1, 0\rangle$), the molecules can undergo energy-conserving spin exchanges mediated by dipolar interactions between nearby lattice sites, as depicted in Fig. 5.1.

© Springer Nature Switzerland AG 2018

J. P. Covey, *Enhanced Optical and Electric Manipulation of a Quantum Gas of KRb Molecules*, Springer Theses, https://doi.org/10.1007/978-3-319-98107-9_5

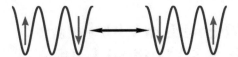

Fig. 5.1 Spin exchange between molecules in $N = 0$ ($\downarrow$) and $N = 1$ ($\uparrow$). Because of the long-range interactions, the exchange can occur over distances of many sites in the lattice

The full spin-1/2 Hamiltonian that describes this exchange process ($J_\perp$) as well as electric field-dependent processes (J_z and W) is [5, 12, 14]:

$$\hat{H} = \frac{1}{2} \sum_{i \neq j} V_{dd}(\mathbf{r_i} - \mathbf{r_j}) \left(\frac{J_\perp}{2} \left(\hat{S}_i^+ \hat{S}_j^- + \hat{S}_i^- \hat{S}_j^+ \right) + J_z \hat{S}_i^z \hat{S}_j^z + W \left(\hat{S}_i^z n_j + \hat{S}_j^z n_i \right) \right),$$

(5.1)

where $V_{dd}(\mathbf{r_i} - \mathbf{r_j}) = \frac{1 - 3\cos^2 \theta_{ij}}{r_{ij}^3}$, which is a geometrical factor pertaining to dipole–dipole interactions, n_i is the population on site i, and $\hat{S}^+$ and $\hat{S}^-$ are spin-1/2 raising and lowering operators. Note that there is another term which depends only on the densities and not on the spins [12]. The coupling constant $J_\perp = \frac{k d_{\downarrow\uparrow}^2}{4\pi \epsilon_0 a_{\text{lat}}^3}$, where $d_{\downarrow\uparrow}$ is the transition dipole moment between the two spin states $| \downarrow \rangle$ and $| \uparrow \rangle$ (see Ref. [12]) (at zero DC field, $d_{\downarrow\uparrow} = \mathcal{D}/\sqrt{3}$), ϵ_0 is the permittivity of free space, a_{lat} is the lattice spacing, and subscripts i and j index lattice sites (Fig. 5.2).

It is constructive to understand the four prefactors and how they depend on the electric field. As stated above, $J_\perp \sim d_{\downarrow\uparrow}^2$, where again $d_{\downarrow\uparrow} = \langle \downarrow | \hat{d}_0 | \uparrow \rangle = \langle \uparrow | \hat{d}_0 | \downarrow \rangle$. The second term has $J_z \sim (d_\uparrow - d_\downarrow)^2$, where $d_\downarrow = \langle \downarrow | \hat{d}_0 | \downarrow \rangle$ and $d_\uparrow = \langle \uparrow | \hat{d}_0 | \uparrow \rangle$ [12]. The third term has $W \sim (d_\uparrow^2 - d_\downarrow^2)/2$, and the fourth term ($\sim n_i n_j$) has $V \sim (d_\uparrow^2 + d_\downarrow^2)/4$ [12]. Note that $k = 2$ for $| \uparrow \rangle = |1, 0\rangle$ and $k = -1$ for $| \uparrow \rangle = |1, -1\rangle$, which are set by the matrix elements coupling to $| \downarrow \rangle \equiv |0, 0\rangle$. With molecules distributed in a 3D lattice, the dominant frequency (energy) component present in the Hamiltonian is $|J_\perp/(2h)|$, where h is Planck's constant. The experiments of Refs. [7, 14] were performed at zero DC electric field, so $J_z = W = 0$ [12], and thus the remainder of this chapter focuses only on the spin-exchange term.

5.2   Spin-Echo Ramsey Spectroscopy

The dynamics were probed with a spin-echo Ramsey spectroscopy sequence, which is described below and schematically depicted in Fig. 5.3a. One additional complication in the experiment is that there is a site-to-site differential energy shift due to a residual inhomogeneous light shift arising from the anisotropic

Fig. 5.2 (**a**) and (**b**) Spin exchange between molecules in $N = 0$ ($\downarrow$) and $N = 1$ ($\uparrow$). Because of the long-range interactions, the exchange can occur in principle over distances of many sites in the lattice. (**c**) The geometrical factor associated with the distance and angle between each molecule and the center molecule (green). Reproduced from Ref. [14]

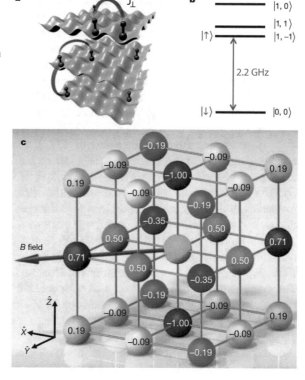

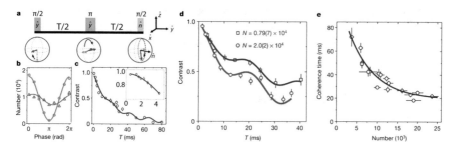

Fig. 5.3 (**a**) Spin-echo setup and timing diagram. (**b**) Typical Ramsey fringes at short time (green) and long time (orange). (**c**) Contrast vs. time showing the decay and oscillations for intermediate density. The inset shows that the contrast decay for short times is concave down. (**d**) Contrast decay for two different particle densities. The oscillation frequency is basically the same for the two datasets but the coherence time is shorter for higher density. (**e**) The data is fit to $C(T) = Ae^{-T/\tau} + B\cos^2(\pi f T)$. Compiling all of the measured coherence times shows $\tau \propto 1/N$. Reproduced with permission from Ref. [14]

polarizability of the molecules, caused by a difference in how the two rotational states couple to the electronic excited states. The use of a "magic" angle (see Fig. 4.9) for the lattice polarization with respect to the quantization axis reduces the inhomogeneity by more than a factor of 10, but a residual effect remains [11]. A spin-echo technique is an effective approach to mitigate the effects of single-particle dephasing from this energy spread across the molecular cloud. Initially, all of the molecules are $| \downarrow \rangle$. The first $\pi/2$ pulse excites every molecule to a coherent superposition of $(| \uparrow \rangle + | \downarrow \rangle)/\sqrt{2}$. After a free evolution time $T/2$, an echo pulse is applied to reverse the single-particle dephasing from the inhomogeneous light shift. After another free evolution time of $T/2$, a final $\pi/2$ pulse is applied with a phase ϕ relative to the first pulse. This rotates the Bloch vector about the axis $\hat{\mathbf{n}} = \cos\phi\,\hat{\mathbf{y}} + \sin\phi\,\hat{\mathbf{x}}$. We measure the number of molecules in $| \downarrow \rangle$ by using STIRAP to transfer the ground-state molecules back to weakly bound Feshbach molecules, followed by absorption imaging of the constituent atoms. By varying the angle ϕ we can obtain a Ramsey fringe contrast for a given free evolution time T, as shown in Fig. 5.3b, which we fit to $\frac{N_{tot}}{2}\left(1 + C\cos(\phi + \phi_0)\right)$, and describes oscillation with amplitude given by the contrast C around 50% the total number. Here, N_{tot} is the total molecule number and C is the fringe contrast ($0 \leq C \leq 1$). The contrast determines the amount of spin coherence left in the system after a time T, and it usually decays as a function of T due to residual single-particle dephasing and many-molecule interaction effects. Some typical contrast decay curves are shown in Fig. 5.3c, d. The most striking feature of these curves is the oscillation superimposed on an overall decay. We attribute both the oscillations and the overall decay to dipolar interactions. Dilute lattice fillings and long-range interactions lead to a spread of interaction energies, which cause dephasing and loss of contrast, and the interaction energy spectrum has the strongest contribution from the nearest-neighbor interaction (of frequency $f = J_\perp/2h$), which we fit empirically to $(1 - A)e^{-T/t} + A\cos^2(\pi f T)$, where $A \sim 0.1$ [7, 14].

5.3   Removing Pairwise Entanglement: WAHUHA NMR Sequence

Figure 5.4a shows how the spin-echo sequence can be extended with a multipulse sequence which removes pairwise entanglement. Such multipulse sequences are examples of dynamical decoupling, which is widely used in nuclear magnetic resonance (NMR) [13] and quantum information processing [9] to remove dephasing and extend coherence times. This sequence in particular removes dephasing due to two-particle dipolar interactions by swapping the eigenstates of the dipolar interaction Hamiltonian for two isolated particles to allow for subsequent rephasing. This pulse sequence is named WAHUHA after its authors—Waugh, Huber, and Haeberlen [13].

Fig. 5.4 The WAHUHA pulse sequence. (**a**) shows the pulse sequence used to remove pairwise entanglement. The first, middle, and last pulse are the same as the spin-echo sequence described above, and the extra pulses serve as an echo sequence in the two-molecule basis. (**b**) shows the contrast decay vs. time for Ramsey (purple), spin echo (red), and WAHUHA (black). The inset shows the difference between the spin echo and the WAHUHA for each time. Reproduced from Ref. [14]

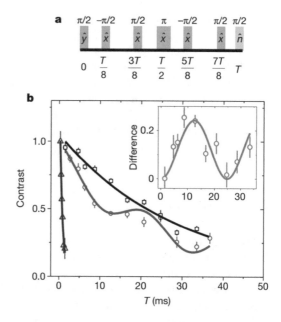

This procedure is straightforward to understand by considering two particles initially prepared in $|\downarrow\downarrow\rangle$. Then an initial $(\pi/2)_y$ pulse transfers them to

$$\frac{1}{\sqrt{2}}(|\downarrow\rangle+|\uparrow\rangle)\otimes\frac{1}{\sqrt{2}}(|\downarrow\rangle+|\uparrow\rangle)=\frac{1}{2}(|\downarrow\downarrow\rangle+|\uparrow\uparrow\rangle+|\downarrow\uparrow\rangle+|\uparrow\downarrow\rangle). \quad (5.2)$$

Because of the spin-exchange term, $|\downarrow\uparrow\rangle$ and $|\uparrow\downarrow\rangle$ are not eigenstates of the Hamiltonian. However, the three triplet states $|\downarrow\downarrow\rangle$, $|\uparrow\uparrow\rangle$, and $(|\downarrow\uparrow\rangle+|\uparrow\downarrow\rangle)/\sqrt{2}$ are eigenstates with eigenenergies 0, 0, and $J_\perp/2$, respectively. Note that a single $(\pi/2)_x$ pulse can swap the states $|\downarrow\downarrow\rangle+|\uparrow\uparrow\rangle$ and $|\downarrow\uparrow\rangle+|\uparrow\downarrow\rangle$, and can thus act as an effective spin echo for the two-particle wavefunction.

During the first free evolution time of duration $T/8$ as shown in Fig. 5.4a, the eigenstates $|\downarrow\downarrow\rangle$ and $|\uparrow\uparrow\rangle$ accumulate no phase, whereas $(|\downarrow\uparrow\rangle+|\uparrow\downarrow\rangle)/\sqrt{2}$ acquires a phase of $e^{-i(J_\perp/\hbar)T/16}$, which entangles the state. After this evolution time we apply a $(-\pi/2)_x$ pulse to swap the contributions from $|\downarrow\uparrow\rangle+|\uparrow\downarrow\rangle$ and $|\downarrow\downarrow\rangle+|\uparrow\uparrow\rangle$, and swapping the accrued phases. After another evolution time of $T/4$, the $(\pi/2)_x$ pulse swaps the phases again. This state then evolves freely for another time $T/8$, after which both $|\downarrow\uparrow\rangle+|\uparrow\downarrow\rangle$ and $|\downarrow\downarrow\rangle+|\uparrow\uparrow\rangle$ have accumulated the same total phase, $e^{-i(J_\perp/\hbar)T/8}$, and the state is no longer entangled as a result. Hence the oscillations due to pairwise dipole–dipole interactions are cancelled.

Figure 5.4b shows the decay of the Ramsey contrast for three different pulse sequences. In purple is the simple two-pulse Ramsey sequence without an echo. The coherence time is ≈ 1 ms, which matches the ≈ 1 kHz energy shift across the cloud discussed in Chap. 4. This is limited by the anisotropic polarizability, hyperpolarizability, and any residual magnetic or electric field noise. In red is shown the three-pulse spin- echo sequence, as discussed in the previous section. The black curve shows the seven-pulse WAHUHA sequence, and the pairwise oscillations have clearly vanished. To elucidate this effect further, the inset shows the difference between the spin-echo and WAHUHA contrasts at each time. We attribute the finite decay time of the coherence that remains after pairwise entanglement removal to many-body (beyond two-body) effects, as well as residual decoherence mechanisms.

5.4 Many-Body Dynamics

While this NMR sequence allows us to remove pairwise entanglement, extending it to larger clusters becomes highly nontrivial. Therefore, to assess the many-body nature of our interactions we must turn to more indirect methods of analysis. The first method is by comparing the dynamics observed in our experiment to a theoretical simulation carried out in the group of Ana Maria Rey. The dotted black lines in Fig. 5.5 (and the lines in Fig. 5.7) are a many-body simulation where the only fitting parameter is the filling fraction. This simulation technique is described in Ref. [7], and represents a step forward in the simulation of complex, many-body systems. Further, this theory is instructive for understanding the experiment.

Specifically, only nearest-neighbor, or nearest-neighbor and next-nearest-neighbor interactions can be considered, as shown in Fig. 5.5a and b, respectively. Again, the only fitting parameter is the filling fraction. The blue and purple lines show these simulations for a few different filling fractions that describe well either the short time or long time part of the contrast decay. However, all of these curves are in qualitative disagreement with the observation and the full theory (black

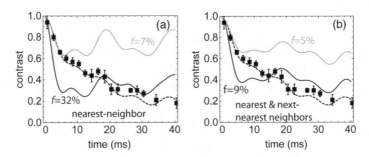

Fig. 5.5 The many-body nature of the dipolar interactions. (**a**) shows the simulations of the spin-exchange dynamics allowing only nearest-neighbor interactions (blue, purple), and the full simulation with the filling fraction as a fitting parameter. (**b**) shows nearest and next-nearest neighbors (blue, purple), with the same simulation in black. Reproduced from Ref. [7]

dotted line), which illustrates that beyond-neighbor interactions are required, and the system is a truly long-range, many-body system.

Another approach for studying the many-body nature of our data is with the simple, exponential fit discussed above: $(1 - A)e^{-T/t} + A\cos^2(\pi f T)$. Looking at the interaction map in Fig. 5.2c, it is clear that there are three dominant interactions with relative frequencies f, $f/\sqrt{2}$, and $f/2$. Therefore, we can consider instead a fit that includes these three frequencies:

$$C(T) = (1 - A)e^{-T/t} + A\left(\cos^2(\pi f T) + \cos^2(\pi(f/\sqrt{2})T) + \cos^2(\pi(f/2)T)\right).$$
(5.3)

Note that we force the amplitude of the three oscillation terms to be identical. While this equation has more terms, it does not have any more fitting parameters, which are still just the frequency f, the coherence time t, and the amplitude A.

Figure 5.6a shows a contrast decay dataset, and it is fit with the one-frequency function (green) and the three-frequency function (blue). The three-frequency fit is clearly better even by eye, but we quantify this in Fig. 5.6b by considering the reduced χ^2 of the two fits. We study this fit quality as a function of frequency, and we find that the three-frequency fit is clearly better near the correct frequency $(J_{\perp,|1,0\rangle}/2h = 100\,\text{Hz})$, and that the three-body fit is best very close to the correct frequency at 100 Hz. This suggests that every molecule is interacting strongly with many of its neighbors, and thus the system is necessarily described by many-body physics (Fig. 5.7).

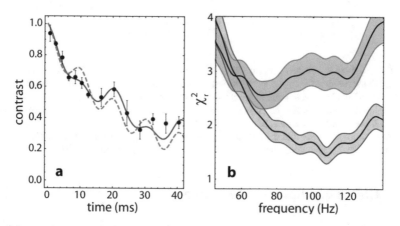

Fig. 5.6 Fitting the contrast decay with three frequencies vs. one frequency for the $|0, 0\rangle$ to $|1, 0\rangle$ transition. (**a**) shows the fitting with one frequency in green, and three frequencies in blue. There is only one fitting parameter f in both cases, but the other frequencies used are $f/\sqrt{2}$ and $f/2$. (**b**) shows the reduced χ^2 associated with both fits vs. f. The three-frequency fit is clearly better. Reproduced from Ref. [7]

Fig. 5.7 Different excited
states to use for
spin–exchange interactions.
The strength of the
spin-exchange term varies by
a factor of two between the
two projections of the excited
rotational state $|1, -1\rangle$ and
$|1, 0\rangle$. Reproduced from
Ref. [7]

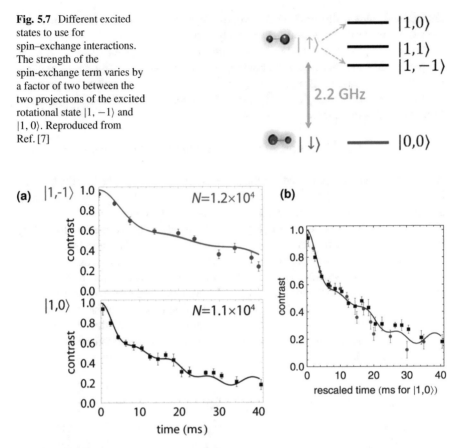

Fig. 5.8 (a) Typical contrast decay curves for $|\uparrow\rangle = |1, -1\rangle$ (top) and $|\uparrow\rangle = |1, 0\rangle$ (bottom) for roughly the same density ($\sim 1.2 \times 10^4$ molecules). The coherence time is clearly shorter for the $|1, 0\rangle$ data, which is expected due to the stronger interactions. (b) Taking the two datasets from (a) and scaling the time axis for the $|1, -1\rangle$ data by a factor of two, we see that the curves collapse onto each other reasonably well, which highlights that dipolar interactions are responsible for the observed dynamics. Note that the data for $|1, -1\rangle$ is rescaled by a factor of two shorter in (b), so data out to 70 ms is included. The lines are theory [7]. Reproduced from Ref. [7]

5.5   Universal Dipolar Interactions

Another way to change the interaction strength is to couple $|0, 0\rangle$ to a different state in the $N = 1$ manifold, as $J_\perp$ is twice as large for the $\{|0, 0\rangle, |1, 0\rangle\}$ transition. To test this, the same experiment described above was repeated with $|\uparrow\rangle = |1, 0\rangle$ [7]. Figure 5.8a shows typical contrast decay curves (coherence time is given by $1/e = 0.37$) for $\sim 10^4$ molecules for both choices of $|\uparrow\rangle$, and Fig. 5.8b compares the two curves when time is rescaled by a factor of $1/2$ for the $|1, -1\rangle$ data. The contrast clearly decays faster for $|\uparrow\rangle = |1, 0\rangle$ for which the interactions are stronger, and the fact that the two curves nearly collapse onto each other when time is rescaled by exactly the same ratio of the interaction energy highlights that dipole–dipole interactions give rise to the observed dynamics. The solid curves are based on theoretical simulations obtained from a cluster expansion [7]. A new approach to the cluster expansion technique called a "moving-average" cluster expansion [7] was developed in order to explain our data, and it exemplifies the need for new development of theory tools to help understand the complex many-body quantum systems that are now found in laboratories, such as trapped ion chains, magnetic atoms, and atoms in optical lattices. This theory comparison also informed us that the molecular filling fraction in the lattice was low, at about 5% for the data in Fig. 5.8, which is consistent with the earlier estimation based on the quantum Zeno effect. This has motivated work to increase the filling fraction in the lattice [10], to be discussed in Chap. 6.

5.6 $N = 1$ to $N = 2$ Rotational Transitions

Another way to tune the strength of the dipolar interaction is to use the $N = 1$ to $N = 2$ transition instead of $N = 0$ to $N = 1$. The rotational transitions, and hence the spin exchange process, is an electric dipole transition. Therefore, only states of opposite parity can be coupled. The next transition on the lattice is $N = 1$ to $N = 2$, for which the dipole moment is ~ 0.1–0.2 Debye rather than 0.567 Debye. The exchange energies of this spin qubit are $J_{\perp, N=1-2}/2 = 10$–20 Hz, and almost all signatures of the spin exchange process should be undetectable. This could serve as a useful benchmark to show that the oscillations and many-body dynamics we observe are not somehow a technical artifact of the experiment, and would supplement other evidence like the density-dependence, many-body theory, and the WAHUHA sequence.

As discussed already in the previous section, reducing the interaction strength increases the coherence time since many-body entanglement is playing less of a role. Thus, for the $N = 1$ to $N = 2$ transition the coherence time should be significantly longer than the $|1, -1\rangle$ case. We measure the transition frequency to be 4.4558 GHz, which is consistent with the rotational constant. We were never able to use this transition for spin-exchange dynamics, however, because we could not find any magic angle to minimize the differential polarizability. We worked on this only briefly, but the magic angle is extremely important for the long coherence times observed with the $N = 0$ to $N = 1$ transitions.

5.7   Pulse Sequences with Variable Tipping Angles

All the pulse sequences shown so far have used pulses with an area of either $\pi/2$ or π, but a lot of information could potentially be gleaned from using other tipping angles [8]. We can write

$$\vec{S}_i \cdot \vec{S}_j = \hat{S}_i^x \hat{S}_j^x + \hat{S}_i^y \hat{S}_j^y + \hat{S}_i^z \hat{S}_j^z. \tag{5.4}$$

In the mean-field picture, the spin operators can be replaced by ensemble averages, and so $\vec{S}_i \cdot \vec{S}_j \approx \langle S \rangle^2$ at short times, provided the initial state can be represented in the Dicke manifold. Then, the spin exchange term can be rewritten $\hat{S}_i^x \hat{S}_j^x + \hat{S}_i^y \hat{S}_j^y \approx \langle S \rangle^2 - \hat{S}_i^z \hat{S}_j^z$. Further, the projection operators can also be replaced with ensemble averages, and so $\hat{S}_i^z$ is replaced by the average magnetization $\langle S^z \rangle$.

The first pulse of our Ramsey sequence of area θ then changes the average magnetization by $\langle S^z \rangle = \cos(\theta)$. Now the spin-exchange term can be thought of in the mean-field picture as a spin interacting with the effective magnetic field due to all the other spins. In this picture the spins accrue a phase shift of $\Delta \phi \sim \cos(\theta) \Delta \tau$, such that the Ramsey fringe evolves in time as $\mathrm{Re}(e^{-i\cos(\theta)\tau})$. This trend is shown in Fig. 5.9a for $\theta = \pi/16$ and $15\pi/16$. Figure 5.9b shows the contrast of the final Ramsey fringe as a function of the hold time, and Fig. 5.9c shows the initial slope from Fig. 5.9a as a function of tipping angle (or pulse area) between zero and π. The corresponding calculations are shown in Fig. 5.9d–f, and they show excellent qualitative agreement with the data.

However, as seen in comparing Fig. 5.9a, d, the amplitude of the data is far too large. In fact, we found that developing good quantitative agreement was incredibly difficult, and we attributed this to the non-magicness of the trap. Due to the anisotropic polarizability [11] and hyperpolarizability [14] discussed in the previous chapter, there is a differential light shift across the cloud. Thus, the tipping angle varies across the cloud, and the spins start to become off-resonant compared to $J_\perp/2$ towards the outside of the cloud. Thus the phase evolution is highly nontrivial to understand, and phase evolution can be observed even at a tipping angle of $\pi/2$ (see Fig. 5.3b) where it should not happen. Many-body effects would certainly cause our experimental observations to differ from the simple mean-field approximation, but we were unable to demonstrate a quantitative agreement between our data and the many-body theory described in the sections above.

This differential light shift that dephases individual spins can be included into the Hamiltonian, and should be written as $1/2 \sum_i h_i \hat{S}_i^z$, which looks like a magnetic field. The strength of h_i varies from site to site in accordance with the harmonic confinement. While this effect can be mitigated with the spin-echo pulse, imperfections become particularly acute at small tipping angles. Therefore, we were never able to learn much about the many-body dynamics of our system by varying the tipping angle, and a better approach to combat the hyperpolarizability would be needed to proceed with these experiments. In the future we would like to make the lattice very flat and have molecules occupy only the central region.

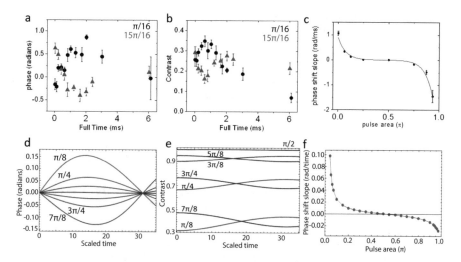

Fig. 5.9 Pulse sequences with variable tipping angle. (**a**) shows the phase of the Ramsey fringe as a function of the dark time when the tipping angle is $\pi/16$ and $15\pi/16$. (**b**) shows the contrast of these Ramsey fringes. (**c**) shows the slope of the initial phase change (e.g., short times in (**a**)) for different tipping angles. (**d**)–(**f**) show simulations of the plots in (**a**)–(**c**). Note that the amplitude in (**d**) is significantly smaller than the measured amplitude in (**a**)

5.8   Inferring the Molecule Filling Fraction

The curves in Fig. 5.3d display a clear density dependence, which is a signature of an interaction effect. The density was varied by loading the same initial distribution of molecules and then holding them in the lattice for a variable amount of time to allow for single-particle loss from light scattering. In Ref. [4], we acquired a full understanding of the inelastic light scattering from the trapping light, and so we can use the trapping light to remove molecules or drive them to dark states at a well-understood rate, which is very long compared to the Ramsey spectroscopy sequence. Thus, the density ρ is proportional to the number of molecules we detect [4]. We expect the coherence time τ to scale with the molecule number N as $\tau \propto 1/N$, since

$$\tau \propto \frac{1}{\langle E_{\text{int}} \rangle} \propto \frac{\bar{R}^3}{J_\perp} \propto \frac{1}{J_\perp \rho}, \tag{5.5}$$

where $\bar{R}$ is the average interparticle spacing and the density $\rho = \bar{R}^{-3}$. For our loading scheme, $\rho \propto N$. Figure 5.3e shows the coherence time clearly scales as $1/N$.

As stated in several places so far in this thesis, the filling fraction in the lattice has been measured to be ~5–10% by several methods, including the decay time of the contrast. To proceed with quantum magnetism or out-of-equilibrium many-body dynamics using dipolar interactions such as the spin-exchange or Ising interactions, it will be important to increase this filling fraction. We have tried many times over the past five or 6 years to reduce the temperature and entropy of our molecular sample by many different methods, and we succeeded in doing so within the last few years using a quantum synthesis approach with atomic insulators in the optical lattice. These efforts are the subject of Chap. 6.

References

1. S. Baier, M.J. Mark, D. Petter, K. Aikawa, L. Chomaz, Z. Cai, M. Baranov, P. Zoller, F. Ferlaino, Extended bose-hubbard models with ultracold magnetic atoms. Science **352**(6282), 201–205 (2016)
2. R. Barnett, D. Petrov, M. Lukin, E. Demler, Quantum magnetism with multicomponent dipolar molecules in an optical lattice. Phys. Rev. Lett. **96**, 190401 (2006)
3. D. Barredo, S. de Léséleuc, V. Lienhard, T. Lahaye, A. Browaeys, An atom-by-atom assembler of defect-free arbitrary two-dimensional atomic arrays. Science **354**(6315), 1021–1023 (2016)
4. A. Chotia, B. Neyenhuis, S.A. Moses, B. Yan, J.P. Covey, M. Foss-Feig, A.M. Rey, D.S. Jin, J. Ye, Long-lived dipolar molecules and feshbach molecules in a 3d optical lattice. Phys. Rev. Lett. **108**, 080405 (2012)
5. A.V. Gorshkov, S.R. Manmana, G. Chen, J. Ye, E. Demler, M.D. Lukin, A.M. Rey, Tunable superfluidity and quantum magnetism with ultracold polar molecules. Phys. Rev. Lett. **107**, 115301 (2011)
6. K.R.A. Hazzard, S.R. Manmana, M. Foss-Feig, A.M. Rey, Far-from-equilibrium quantum magnetism with ultracold polar molecules. Phys. Rev. Lett. **110**, 075301 (2013)
7. K.R.A. Hazzard, B. Gadway, M. Foss-Feig, B. Yan, S.A. Moses, J.P. Covey, N.Y. Yao, M.D. Lukin, J. Ye, D.S. Jin, A.M. Rey, Many-body dynamics of dipolar molecules in an optical lattice. Phys. Rev. Lett. **113**, 195302 (2014)
8. M.J. Martin, M. Bishof, M.D. Swallows, X. Zhang, C. Benko, J. von Stecher, A.V. Gorshkov, A.M. Rey, J. Ye, A quantum many-body spin system in an optical lattice clock. Science **341**(6146), 632–636 (2013)
9. P.C. Maurer, G. Kucsko, C. Latta, L. Jiang, N.Y. Yao, S.D. Bennett, F. Pastawski, D. Hunger, N. Chisholm, M. Markham, D.J. Twitchen, J.I. Cirac, M.D. Lukin, Room-temperature quantum bit memory exceeding one second. Science **336**(6086), 1283–1286 (2012)
10. S.A. Moses, J.P. Covey, M.T. Miecnikowski, B. Yan, B. Gadway, J. Ye, D.S. Jin, Creation of a low-entropy quantum gas of polar molecules in an optical lattice. Science **350**(6261), 659–662 (2015)
11. B. Neyenhuis, B. Yan, S.A. Moses, J.P. Covey, A. Chotia, A. Petrov, S. Kotochigova, J. Ye, D.S. Jin, Anisotropic polarizability of ultracold polar ^{40}K^{87}Rb molecules. Phys. Rev. Lett. **109**, 230403 (2012)
12. M.L. Wall, K.R. A. Hazzard, A.M. Rey, Quantum magnetism with ultracold molecules, in *From Atomic to Mesoscale*, Chap. 1 (World Scientific, Singapore, 2015), pp. 3–37
13. J.S. Waugh, L.M. Huber, U. Haeberlen, Approach to high-resolution NMR in solids. Phys. Rev. Lett. **20**, 180–182 (1968)
14. B. Yan, S.A. Moses, B. Gadway, J.P. Covey, K.R.A. Hazzard, A.M. Rey, D.S. Jin, J. Ye, Observation of dipolar spin-exchange interactions with lattice-confined polar molecules. Nature **501**(7468), 521–525 (2013)

Chapter 6
A Low-Entropy Quantum Gas of Polar Molecules in a 3D Optical Lattice

The first successful creation of KRb ground-state molecules in 2008 reached close to quantum degeneracy, with a temperature of $1.3T_F$. However, despite an intense effort to further cool the molecular gas, the lowest temperature for the KRb gas reached in a harmonic trap is $T/T_F = 1$. No other groups have produced colder, denser gases to date. As discussed at the end of Chap. 5, the best filling fraction we had previously reached in the optical lattice is 5–10%. I will begin this chapter by discussing why it is important to increase the filling fraction, and what new and exciting experiments await a low-entropy molecular sample. Then, I will discuss many of our efforts to reduce the entropy, with the ultimate success coming from a quantum synthesis approach of atomic insulators in the lattice. This successful approach will then be the topic for the remainder of the chapter.

6.1 The Need for Low Entropy: Percolation Theory

Low temperatures and entropies are required for most of the exciting future directions to be pursued with polar molecules. For bulk gases, one would expect qualitatively new phenomena when the temperature is on the order of the dipolar interaction energy. For KRb at moderate densities of $n = 10^{13} \, \mathrm{cm}^{-3}$, the dipolar interaction has an energy scale that corresponds to on average 5–10 nK. Therefore, it will be important to reach temperatures well below 100 nK.

In the lattice, on the other hand, the desire for low entropy comes from a need for a high filling fraction. Nearly, any quantum simulation or quantum information research requires a well-defined initial state, where the location of each particle is known with high fidelity. Indeed, realizing exotic quantum magnetic phases of matter in the XXZ model would likely require near-unity filling. Moreover, since

© Springer Nature Switzerland AG 2018
J. P. Covey, *Enhanced Optical and Electric Manipulation of a Quantum Gas of KRb Molecules*, Springer Theses, https://doi.org/10.1007/978-3-319-98107-9_6

Fig. 6.1 Low-entropy lattice cartoon. This shows a high filling fraction of molecules in the lattice and illustrates how they are all connected in a percolating network. Reproduced from Ref. [22]

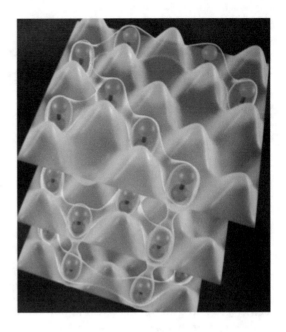

the energy scale between neighboring sites is $J_\perp/2 = 100\,\text{Hz}$, most dynamical experiments centered around the spin–exchange mechanism will require a dense network of nearest-neighbor interactions. While we have demonstrated conclusively that the interactions we observe are many-body, the interaction range is too small to support significant interaction between even next-nearest-neighbor sites. Additionally, the hyperpolarizability discussed in Chap. 4 causes an energy offset between distant sites which adds a detuning in their interactions (Fig. 6.1).

It is constructive to understand what filling fraction is actually needed for the onset of dynamics in an out-of-equilibrium spin system. To do this, we invoke the idea of a percolation threshold, which is the formation of long-range connectivity in a random system. Below the threshold, a large connected component of the system does not exist; while above it, there exists a connected component with a size on the order of the system size. The notion of connectivity is based on the competition between the energy scales discussed above. Since the dipolar interaction has finite range, inhomogeneity or finite coherence time will set the length scale in the system. The hyperpolarizability causes a kHz-level energy shift across the cloud and limits the Ramsey coherence time to 1 ms without an echo pulse. Therefore, even next-nearest-neighbor interactions at $\sim 10\,\text{Hz}$ energy scales are starting to set the length scale of the interaction. Accordingly, percolation occurs when every molecule has a nearest neighbor with which it can interact.

The percolation threshold for an infinite cubic lattice of nearest-neighbor interactions can be calculated analytically, and is given in Ref. [32] to be $f_{\text{PT}} = 0.3$. Since our system has finite size and our interactions are slightly longer than nearest neighbor, we expect the percolation threshold in our system to be <0.3. Therefore, this number serves as a benchmark for what we would like to achieve in our experiment, and we have been trying since 2010 or so to create a lower-entropy system.

6.2 Many Previous Efforts

There are several challenges to efficiently make Feshbach molecules (FbM) in a bulk 3D gas. The first is good spatial overlap between the two clouds. There are two reasons why this could be poor. As the clouds become very cold and the optical trap becomes very weak, the difference in mass causes the two clouds to separate. The differential sag is given by $\Delta y_{sag} = -g(1/\omega_{Rb}^2 - 1/\omega_K^2)$ [24], where ω is the oscillation frequency in the trap. In our optical traps at $\lambda_1 = 1064$ nm, the polarizability of K and Rb are similar to within 80%, and so their oscillation frequencies are different by $\approx \sqrt{m_{Rb}/m_K}$. In typical conditions of a degenerate Bose–Fermi mixture of K and Rb, this differential sag could be $\approx$3–5 μm.

The other problem with their spatial overlap also arises from reaching quantum degeneracy. The fermions have enormous Fermi pressure due to the Pauli exclusion principle. Typical sizes for even the coldest degenerate Fermi gases of $N_K \approx 10^5$ are $\sigma_K \approx 50\,\mu$m. Conversely, the bosons condense to a very small cloud described by the Thomas–Fermi (TF) approximation, and typical conditions are $N_{Rb} \approx 10^5$ where $\sigma_{Rb} \approx 10\,\mu$m. In this regime, the differential sag becomes similar to the size of the Rb cloud. Therefore, not only are their sizes very poorly matched, their positions are significantly different. The latter problem cannot be solved easily with a magnetic field gradient which generates a differential force because of the magnetic moments of the states we use.

In addition to these problems, the high density of the BEC introduces another problem with three-body loss. If even 50% of the Rb is converted to an FbM, the remaining 50% will quickly cause three-body interactions between one FbM and two Rb atoms. The lifetime of FbM with atoms is several ms, which makes it difficult to do STIRAP and remove the unpaired atoms quickly enough. The timescale of three-body loss increases as n_{Rb}^3, so dense BECs pose significant challenges.

The next challenge is the poor momentum space overlap. The BEC has macroscopic population of the lowest momentum state of the trap by definition, but the DFG stacks up K atoms in momentum states with one per level, as required by the Pauli exclusion principle. Therefore, similar to coordinate space, the momentum space distribution of K is much larger than Rb. It is constructive to think about how this pairing looks in momentum space, and a few possibilities are shown in Fig. 6.2. The ideal case is when each Rb takes the lowest K available, such that the momentum distribution of the molecules is the same as K, and the FbM T/T_F is similar to that of K. This would allow us to make a DFG of molecules in bulk, but this is certainly not what actually happens.

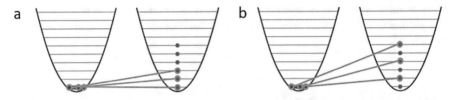

Fig. 6.2 Degenerate K–Rb pairing in momentum space. (**a**) The ideal case where each Rb (red) picks the next lowest available K (blue). In our experiment, this typically occurs when $N_{Rb} \gg N_K$, but then we suffer strong three-body loss. (**b**) The more realistic case where each Rb picks a K at random through the momentum distribution, and is actually biased to the top by the weight of the Fermi surface. In our experiment, this typically occurs when $N_{Rb} \ll N_K$

Realistically, each Rb will pair with any K atom in the distribution, but it will be biased by the Fermi surface which carries the most weight because of the $k^2 dk$ factor in the momentum integral. Therefore, the FbMs actually inherit the temperature of K (or slightly higher) rather than its T/T_F. Moreover, since $N_{FbM} \approx 0.2 N_K$, the T/T_F immediately jumps up by a factor of ≈ 3. Concomitantly, even starting with $T_K = 0.3 T_F^K$, we have never been able to create FbM colder than $T_{FbM} = 1.0 T_F^{KRb}$. Nevertheless, we have pursued virtually every possible route to increasing the phase-space-density, and I will now discuss several of these efforts.

6.2.1   Species-Selective Dipole Traps

Despite the complications with momentum space overlap, we spent a lot of time trying to improve the coordinate space overlap. Since the K cloud is bigger and higher in the trap than Rb, a species-selective trap which only affects K could be very useful. In order to make the polarizability of the trap equal to zero for Rb, we need the contributions from the $P_{3/2}$ (called D2) at 780 nm and $P_{1/2}$ (called D1) at 795 nm to cancel each other. This is done at a wavelength of 790 nm, though it is sensitive to the polarization since vector and tensor light shifts are significantly so close to resonance.

As discussed in Chap. 4, this trap can be used to move the K cloud relative to Rb (see Fig. 4.14), and it can be used to compress K. We had to be careful, however, to limit the heating of K and Rb, since the scattering rates are high for both and the clouds are initially very cold. Nevertheless, this beam was actually very useful since it allowed us to efficiently make FbM at very low temperatures in very weak traps, and we used this approach to make $\approx 1 - 2 \times 10^4$ FbM at $T = 70$ nK ($T/T_F = 1$). This is the coldest molecular sample ever created as far as I am aware.

The caveat is that this beam kills FbM in 1 ms and ground-state molecules (GSM) in 100 ms, as shown in Fig. 4.14. Therefore, as soon as FbMs are created, we had to turn the 790 nm beam off suddenly. In doing so, the FbMs are out of equilibrium since the trap was changed adiabatically, and thus they slosh and breathe in the trap with large amplitudes. Further, since the lifetime of FbM is only ≈ 15 ms even in ideal circumstances, the FbM must be converted to GSM immediately. At these temperatures, molecules cannot collide with s-wave collisions and hence cannot thermalize. Consequently, the initial improvement in the phase-space-density cannot be utilized, and we ultimately moved on to other approaches.

6.2.2   'Magic' Traps for Matching the Oscillation Frequencies

A related idea which we also pursued is a "magic" wavelength red-detuned trap, where the polarizability ratio matches the mass ratio in such a way that the trap frequencies are equivalent. Specifically, we want $\alpha_K = m_K \cdot \alpha_{Rb}/m_{Rb}$. This occurs for K–Rb at 807–808 nm, depending on the polarization. Such a trap was used extensively in the work of Ref. [24] to remove the differential gravitational sag. However, as with 790 nm, it is important to minimize heating due to off-resonant light scattering.

Though we did not study the effects of this beam too thoroughly, we learned that the FbM and GSM also suffered from short lifetimes. Thus, it was difficult to take advantage of this trap, and typically we found that the benefit from reducing sag was not offset by the heating of the atomic clouds and the fast loss of the molecular samples. However, if we were to load atoms into the lattice with the aid of the 808 nm beam, and then turned it off before making molecules, this beam could be very useful. Indeed, we may return to the idea of a magic trap later, but we will always have to be careful to avoid the detrimental effects of the atomic heating.

6.2.3   Sympathetic Cooling of KRb with Rb

With limited success in creating a degenerate bulk gas of molecules in 3D simply through high-fidelity magneto-association, we now turn to methods by which the molecules could be cooled. Perhaps the most obvious approach to cooling a gas of molecules is to sympathetically cool it with an atomic species, such as Rb. When we make molecules, there are many leftover Rb atoms. By evaporatively cooling the remaining Rb and allowing it to thermalize with KRb, it could be possible to further cool KRb into quantum degeneracy. However, this requires a suitable elastic cross section for Rb–KRb collisions, and a sufficiently small inelastic collision cross section.

While KRb was calculated to be unstable (exothermic) with K and a lifetime of only several ms was observed [23], KRb is expected to be stable (endothermic) with Rb [23]. Despite these calculations, the lifetime of KRb with Rb was measured to be only 60 ms, and the Rb density was quite low, $n_{Rb} = 0.6 \times 10^{12}$ cm^{-3} [23]. To reach even this lifetime, however, it is essential to transfer KRb to its lowest energy hyperfine state, while Rb was similarly in its lowest hyperfine state $|1, 1\rangle$. The short lifetime was attributed to a large fraction of dark molecules which originated from FbM that underwent vibrational-quenching collisions with Rb. If this was the case, then sympathetic cooling with Rb would be very difficult since we cannot spend less time in the FbM step.

Nevertheless, we set out to test the hypothesis that large numbers of KRb molecules in dark states are created during the FbM step by collisions with Rb. We did this by using a lattice to prevent collisions between Rb and FbM. We loaded K and Rb into the lattice, and converted sites with one of each into an FbM with some fidelity. Then, the FbMs were converted to GSM with ∼90% fidelity. The unpaired K was then removed with a blast pulse of resonant light. This light would also dissociate any weakly bound dark molecules (such as d-wave FbM, to be discussed later), and the resulting K would then be removed. Then, the lattice should only contain sites with GSM, a small number of deeply bound dark molecules, and Rb. Next, the GSM are transferred to the lowest energy hyperfine state with ≈95% fidelity. Finally, the lattice can be ramped down, and the lifetime of GSM with Rb can be probed in a cleaner system.

Upon doing this experiment, we found that the lifetime is $\tau = 60$–100 ms, which is similar to the original measurement without the lattice. We can verify that it is indeed essential for KRb to be in the hyperfine ground state, and the lifetime is $\tau < 10$ ms if we leave it in its original state. These results suggest that Rb is indeed the cause for the loss. Reference [23] found a very weak dependence on the atomic density, which is inconsistent with three-body Rb–Rb–KRb collisions.

We believe that this loss is due to the sticky collisions discussed in Chap. 4. An Rb atom collides with a KRb molecule and forms a collision complex where they spend a relatively long time near each other (long compared to the collision rate). Although they would not undergo chemistry, they spend enough time together for a second Rb to hit their complex. The second Rb drives the Rb–KRb complex to a deeply bound state and carries away the binding energy as kinetic energy. Because this process does not require all three particles to collide at the same time, its dependence on density is very shallow. Thus, sympathetic cooling of KRb with Rb does not appear to be an option for cooling the molecules.

6.2.4   Evaporative Cooling

The only remaining option to cool the molecules is with direct evaporative cooling. As discussed in Chap. 3, ultracold fermionic spin-polarized molecules cannot collide in the s-wave channel, as is often used for thermalization during evaporation. Moreover, KRb molecules undergo large inelastic losses due to chemistry, and evaporation will require a large ratio of elastic to inelastic collisions. Therefore, upon first inspection this approach seems very unlikely to succeed. In fact, the solution will be to apply a large electric field.

The dipolar interaction mixes the partial waves, which allows thermalization of identical fermions even at ultralow temperatures. The dipole–dipole interactions

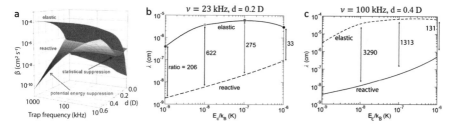

Fig. 6.3 Elastic and inelastic collision rates in 2D. (**a**) The 2D elastic (red) and inelastic, or reactive (yellow/green/blue) collision rates as a function of the dipole moment and the axial trapping frequency which confines in 2D, for a collision energy of 500 nK. Reproduced from Ref. [25]. (**b**) The same elastic and reactive collision rates as a function of collision energy, specifically for $v = 23$ kHz and $d = 0.2$ D. These conditions are appropriate for our efforts with the first-generation apparatus. (**c**) The elastic and reactive collision rates as a function of collision energy for $v = 100$ kHz and $d = 0.4$ D. These conditions are appropriate for what we anticipate doing in the new apparatus

prevent the elastic cross section between identical fermions from vanishing at $T = 0$ because the corresponding Wigner threshold law gives a finite energy-independent elastic cross section [1]. This feature has been used in evaporation of highly magnetic atoms, enabling very efficient evaporation of spin-polarized fermionic samples down to $T < 0.2T_F$ [1, 21]. The dipolar interactions for molecules are much stronger than for highly magnetic atoms, and therefore we expect this approach also to work for us.

As discussed in Chap. 3, molecules can be stabilized in 2D using an electric field that is oriented perpendicular to the plane, such that their collisions are aligned only with the repulsive part of the anisotropic dipolar interaction. This suppresses chemical reactions, and we were able to observe lifetimes of ≈ 1 s at $d = 0.2$ D [8]. For evaporation, an important figure of merit is the ratio of elastic collisions which can thermalize the sample, to inelastic collisions which lead to chemistry and loss. Figure 6.3a shows both the elastic collision rate and the reactive collision rate as a function of both the dipole moment and the axial trapping frequency in the 1D lattice. A larger dipole moment increases the strength of the repulsive dipolar interactions, further stabilizing the molecules. On the other hand, as the axial trap frequency is reduced, the ability for the molecules to collide head-to-tail increases, and thus the reactive loss increases. Therefore, the electric field needs to be very large and the axial trap frequency needs to be very high for successful evaporation.

For the experiments done in 2011 and 2012 in the first-generation chamber, we could only reach a field of 4 kV/cm, which corresponds to 0.2 D. Moreover, at the time we could only reach an axial trap frequency of $v = 23$ kHz. Figure 6.3b considers this exact case as a function of the collision energy, e.g., temperature. At the time, we were starting at a temperature of $T \approx 500$ nK, where the ratio of elastic to inelastic collisions was roughly 100. This ratio is marginally sufficient to successfully evaporate and increase the phase-space-density [39], and so we attempted evaporation.

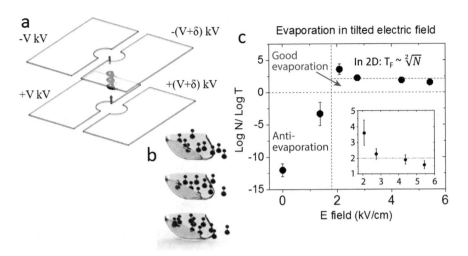

Fig. 6.4 Evaporative cooling of molecules in 1D lattice using an electric field gradient. (**a**) Split plate electrodes are placed around the vacuum cell, and a 1D lattice propagates through a small hole between them. The voltages can be biased to one side to create a radial electric field gradient of a vertical electric field. (**b**) All the molecules on each layer of the 1D lattice are aligned with the electric field. The radial gradient spills the layers and tips them in one direction such that the hottest molecules can be selectively removed. (**c**) The efficiency of the evaporative cooling vs. the electric field strength. Below ∼2 kV/cm, there is anti-evaporation due to the inelastic chemical loss, which happens selectively with the coldest molecules. Above ∼2 kV/cm, there is cooling, and the phase-space-density is increasing above 4 kV/cm. The inset zooms in on this region, and the evaporation efficiency is actually becoming quite good up to the limit of the electric field we can reach in the old apparatus

Figure 6.4 summarizes our evaporation efforts. We used ITO-coated plates to generate an electric field in the vertical direction, and the plates had a small hole in the middle to accommodate a vertical optical lattice (see Fig. 6.4a). The coating on the plates was split into half across the center such that four unique voltages can be applied. A small bias can be added to the two plate sections on the right, which creates an electric field gradient in the radial (horizontal) direction. This gradient tilts the individual 2D traps, and spills the hottest molecules as depicted in Fig. 6.4b.

To characterize the evaporation, we plot the slope of a line fitted to data on a $\log(N)$ versus $\log(T)$ plot, which we label as $\log(N)/\log(T)$ in Fig. 6.4c. With no electric field to stabilize the dipoles, reactive loss causes anti-evaporation since chemistry preferentially occurs in the coldest, highest-density part of the cloud. Therefore, $\log(N)/\log(T) < 0$ since the cloud gets hotter as the loss occurs. At about 2 kV/cm, the electric field is sufficiently high to stabilize the dipoles enough to prevent anti-evaporation, and thus $\log(N)/\log(T)$ crosses zero. Now, let us consider what slope sets the criteria for evaporation that increases the phase-space-density of the molecules.

The Fermi temperature in a 2D gas is given by $T_F = \sqrt{N}\hbar\omega_r/k_B$, where ω_r is the radial trap frequency [10]. To increase the phase-space-density, we want to reduce T/T_F, and thus we want T to decrease faster than $\sqrt{N}\hbar\omega_r/k_B$. Specifically, we want to look at $\log(T)$ versus $\log(T_F) = 1/2\log(N) + \log(\hbar\omega_r/k_B)$. Therefore, in order to increase the phase-space-density we require $\log(N)/\log(T) < 2$. Figure 6.4c shows that the phase-space-density is indeed increasing at fields above $\approx 4.5\,\mathrm{kV/cm}$. However, we were never able to evaporate enough to make a degenerate gas with $T/T_F < 0.5$.

The procedure for initializing this experiment was very detrimental to its success for the following reason. We created molecules in a bulk 3D gas, and then ramped up the 1D lattice and turned on the electric field. As a result, any non-adiabaticity during the ramp would heat the molecules, which cannot thermalize until the electric field is applied. Moreover, since the cloud is initially in a 3D trap, the 1D lattice slices the cloud into ≈ 10–15 pancakes [8]. The number of atoms in the center pancakes is only ≈ 4–5000, coming from the initial cloud of $\approx 3 \times 10^4$. This fact immediately increases T/T_F by a factor of three or so. Moreover, the density increases when we go to 2D, and so there is adiabatic heating of the molecules. While adiabatic heating does not affect T/T_F, higher temperatures cause the ratio of elastic to inelastic collisions to decrease.

Additionally, T_F changes in going from 3D to 2D. In 3D, $T_F^{3D} = \hbar\bar\omega/k_B \cdot (6N)^{1/3}$ [26], while in 2D $T_F^{2D} = \sqrt{N}\hbar\omega_r/k_B$. $\bar\omega = \sqrt{\omega_r^2\omega_z}$ is the geometric mean trap frequency, where $\omega_r \approx 2\pi \times 40\,\mathrm{Hz}$ and $\omega_z \approx 2\pi \times 250\,\mathrm{Hz}$. Note that ω_z does not enter into T_F^{2D}, and therefore T/T_F^{2D} takes another hit to an even higher value since $\omega_z > \omega_r$. The different scaling with number is in the favor of 2D, but both scalings are so shallow that the number change going from 3D to 2D dominates over the change in scaling. As a result of all of these factors, we typically start evaporation with $T/T_F^{2D} = 4$ even though we started in 3D with $T/T_F^{3D} = 1$. The best evaporation we have observed allowed us to get back to $T/T_F^{2D} \approx 1$–2.

During all of these studies, we also had enormous problems with charging of the glass cell and the resulting electric field instability, as discussed in Chap. 2. The new apparatus is designed to reach much larger electric fields, and since the electrodes are in the chamber, we anticipate excellent field stability. This will be discussed more in Chap. 7. The ability to reach 20–30 kV/cm will give us dipole moments of $d = 0.4$ D. Moreover, we expect to be able to reach much larger lattice depths, providing axial trapping frequencies of $\nu = 100\,\mathrm{kHz}$. These conditions are shown in Fig. 6.3c, and the ratio of elastic to reactive collisions will be roughly 400, which is $\approx 4\times$ better than the conditions used in the old apparatus. Moreover, as the molecules get colder this ratio increases very rapidly, even to 3000 at 10 nK.

Another important improvement can be made in the new apparatus by starting in a 2D (or 2D-ish) trap from the beginning. This could be done by loading atoms into the lattice and performing atom evaporation to reach degeneracy in 2D before making molecules on each pancake. Alternatively, the 3D trap aspect ratio could be increased such that the cloud in 3D would load into only a few pancakes. While the prospects for evaporation in the new apparatus are very exciting, this list of problems with our efforts in the old apparatus has made evaporative cooling of molecules to quantum degeneracy prohibitively difficult in the old apparatus. We will return to this experiment in the new apparatus, as discussed in Chap. 10.

6.3   Quantum Synthesis with Atomic Insulators in the Lattice

Thus, it gradually became more attractive to create KRb molecules directly in a 3D optical lattice and optimize the filling fraction to realize a low-entropy system. A natural way to accomplish this task is to take advantage of the precise experimental control that is available for manipulating the initial atomic quantum gas mixture in the 3D lattice. Specifically, we need to prepare low-entropy states of both atomic species and then utilize the already familiar techniques of coherent association and state transfer for efficient molecule production at individual lattice sites (see Fig. 6.5). The combination of efficient magneto-association of preformed pairs [4] and coherent optical state transfer via STIRAP means that the second step should work well; however, creating a low-entropy state for both species with the optimal density of one particle each per site is very challenging. This idea was first proposed in 2003 [6] and later specifically for the KRb system [11]. The creation of Rb_2 Feshbach molecules in a 3D lattice followed this basic idea, where the molecules were produced out of the region of a Mott insulator that has two atoms per site [34]. However, the KRb system faces a much bigger challenge due to the fact that we must address two different species with different masses and different quantum statistics.

The initial experimental target is to produce spatially overlapped atomic distributions that consist of a Mott insulator (MI) of Rb bosons [13] and a spin-polarized band insulator of ^{40}K fermions [29]. The occupancy of the MI depends on the ratio of the chemical potential μ to on-site interaction energy U, and can be higher than one per site if there are too many Rb atoms or if the underlying harmonic confinement of the lattice is too high. The filling fraction of fermions in the optical

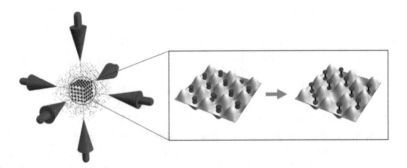

Fig. 6.5 Quantum synthesis of a low-entropy molecular gas in the optical lattice. The ^{40}K Fermi gas needed to obtain a band insulator is much larger than an Rb BEC which gives one atom per site, and the goal is to efficiently make molecules in the center where there is a ^{40}K atom and an Rb atom at each site. Reproduced from Ref. [22]

lattice depends on an interplay between initial temperature, lattice tunneling, and external harmonic confinement. For our experiment, we need a sufficiently tight harmonic confinement and large ^{40}K atom numbers to achieve a filling approaching unity in the center of the lattice. The main challenge of the experiment is the opposite requirements between the filling of the Fermi gas and the number of Rb atoms that can be accommodated in the $n = 1$ Mott insulator. This requires a careful compromise to achieve experimental optimization.

6.4   Experimental Considerations for In Situ Imaging

In order to measure the filling fractions of K and Rb by in situ detection in the lattice, there are a number of crucial experimental systematic checks that must be performed. Therefore, before we go on to learn the atom numbers of K and Rb that are required to reach nearly unity peak occupation for both species, we must calibrate the imaging system and remove imaging systematics that could significantly impact the interpretation of the in situ measurements.

6.4.1   Calibrating the Saturation Intensity

For absorption imaging, it is very important to know the intensity of the probe beam since the scattering rate, and thus the OD, will depend on it. The pictures of our cloud are generated using three images: the first uses resonant light to probe the atoms which in turn cast a shadow onto the camera. This image is called I_{shadow}. The second image contains the probe pulse, but (typically 500 ms) after the atoms are released from the trap. This is called I_{light}. Finally, an image without the probe beam at all is taken to subtract the background light. This is called I_{dark}. Therefore, we use $I_f = I_{shadow} - I_{dark}$ and $I_i = I_{light} - I_{dark}$. These calculations are done for every pixel, and a simple estimate of the OD is

$$\mathrm{OD}_{meas} = \ln\frac{I_i}{I_f}. \tag{6.1}$$

However, if any photons cannot be absorbed by the atoms, which we call "bad light," then the measurable OD will be limited accordingly. For all of the measurements reported in this chapter, light with linear polarization is used to drive a σ^+ transition. Therefore, the absorption cross section is reduced by a factor of two. The maximum OD that can be measured is called OD_{sat}, and this is typically measured to be between three and five by simply looking at the cut-off OD of very dense BECs where the OD is roughly 10. We account for this effect with a modified OD given by [36]:

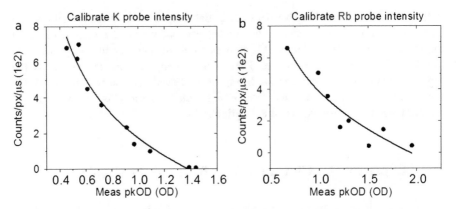

Fig. 6.6 Calibrating the saturation intensity of the probe beam. (**a**) The number of counts on the camera per pixel per µs for the K probe beam as a function of the peak OD of the K cloud that is imaged. (**b**) The number of counts on the camera per pixel per µs for the Rb probe beam as a function of the peak OD of the Rb cloud that is imaged

$$\mathrm{OD}_{mod} = \ln\frac{1 - e^{-\mathrm{OD}_{sat}}}{e^{-\mathrm{OD}_{meas}} - e^{\mathrm{OD}_{sat}}}. \tag{6.2}$$

The intensity of the probe beam plays a significant role, and so the actual OD must account for the intensity, through [36]

$$\mathrm{OD}_{actual} = \mathrm{OD}_{mod} + (1 - e^{-\mathrm{OD}_{mod}})\frac{I}{I_{sat}}. \tag{6.3}$$

In general, the saturation intensity I_{sat} of the probe beam is not known, so this equation can be used to calibrate the intensity. This is done by measuring OD_{mod} (using OD_{meas} and OD_{sat} as described above) as a function of the intensity, as shown for K and Rb in Fig. 6.6. Then, this equation is rearranged to give I as a function of OD_{mod}, with OD_{actual} and I_{sat} as fitting parameters.

6.4.2   Measuring the Imaging Resolution and the Pixel Size

An extremely important parameter for looking at small Mott insulators in the lattice is the resolution, as it artificially inflates the apparent size of small objects. We focus the imaging system by imaging the smallest object we can make reliably (typically a few hundred atoms at a few nK), and then we scan the position of the camera to minimize the size and maximize the optical depth (OD). We measure the resolution at the optimal focus position by measuring the size of the object. We can then vary the size of the BEC loaded into the lattice, and compare the measured size to calculations of the Thomas–Fermi (TF) radius (i.e., when kinetic

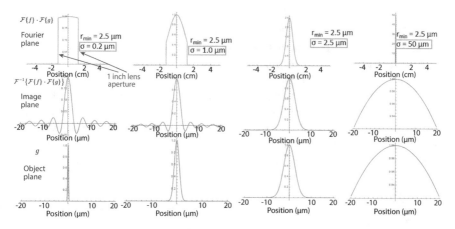

Fig. 6.7 Fourier optics analysis of the imaging system. The first row shows the convolution of the object g with the resolution of the lens f. Then, the Fourier transform is taken to show the transfer function of the lens to the image plane. Using the convolution theorem, this can also be thought of as the Fourier transfer of the cloud times the Fourier transform of the lens. Thus, a small cloud becomes very large in the image plane, and a very large cloud becomes small. The Fourier transform of the lens is a circular aperture, which limits the size in the image plane. From left to right, the cloud size σ is increasing, and thus the limitation of the aperture is decreasing. The next row shows the inverse Fourier transform of the first row, which is done by the second (or the so-called eyepiece lens), and this is what the object looks like on the camera. Note the fringes on the object when the cloud is very small, which serve to form an Airy disk pattern. The third row shows the object function g, which can be compared to the second column. Clearly, the size is limited by the resolution for the objects in the left two columns

energy is neglected) corresponding to the appropriate trap frequencies and atom number calculated from the measurements. The size at which the calculation starts to significantly deviate from the measurement indicates the resolution limit. For typical BECs of Rb, $R_{TF} \approx 10\,\mu$m. Due to the limitations of all the coils in the IP trap around the cell and the cell's poor optical quality, our imaging resolution was limited to $\approx 3\,\mu$m. Therefore, the resolution limit was already playing a significant role for all but the largest BECs that we studied.

It is constructive to use Fourier optics to calculate the effect that finite resolution has on the measured size of the cloud. Consider the cloud to be a Gaussian function g of width σ. The third row of Fig. 6.7 shows the function g for four different values of σ. In generating the image, the object plane is mapped with the first lens to its Fourier transform but is first convolved with the resolution (aperture size) of the lens. This function is given by an Airy function, which is simply a Bessel function, and is labeled as f. Thus, the image plane is given by $F\{f \star g\}$, where F denotes the Fourier transform, and $\star$ denotes the convolution. This can be rewritten using the convolution theorem as:

$$F\{f \star g\} = F\{f\} \cdot F\{g\}. \tag{6.4}$$

The Fourier transform of the Gaussian cloud g is another Gaussian, where the size is inverted. The Fourier transform of the lens Airy function f is given by a disk, or aperture of the lens. The product of these two Fourier transforms is shown in the first row of Fig. 6.7. Note that when σ is very small its Fourier transform is large, and then it is significantly limited by the aperture of the lens.

The second lens of the imaging system images the image plane from the first lens on the camera by taking its Fourier transform. The second row of Fig. 6.7 shows the image on the camera after the second lens. Note that when σ is very small the object on the camera is again small, but limited by the resolution, and the ripples of the Airy function are clearly visible. When σ is very large, the cloud is completely unaffected by the imaging system. This can be seen by comparing to the third row, which shows the cloud function g. This analysis was done for a magnification $M = 1$, but it can easily be generalized to an arbitrary M by simply multiplying the width in the second row by M.

To understand how the imaged size σ_{image} compares to the cloud size σ, we can fit the image function in the second row to a Gaussian. In plotting σ_{image} versus σ, we can generate a rough estimate of how the resolution affects the fitted image size. The resolution function f is given by r_{min}, which is the radius where the Airy function first crosses zero. To convert this to a Gaussian σ_{min}, we must first multiply by ≈ 0.75 to get the $1/e^2$ value (fitting the Airy function to a Gaussian), and then divide by 2 to get the $1/e$ value. Thus, the resolution in the first column $r_{min} = 2.5\,\mu m$ becomes $\sigma_{min} = 1\,\mu m$. Such analysis suggests that:

$$\sigma_{image} = \sqrt{\sigma^2 + \sigma_{min}^2} \tag{6.5}$$

is an excellent estimate for most of the range, though it becomes unreliable when $\sigma \approx \sigma_{min}$. As described above, σ can be calculated from the TF approximation, and this method can be used to measure σ_{min}.

Another effect that was very important in order to avoid artificially inflating the cloud size was the optical repump beam. As discussed in Chap. 3, we use Rb in the $|1, 1\rangle$ state when we make molecules, yet we must image the atoms in the $|2, 2\rangle$ state in order to drive the cycling transition to the $|3, 3\rangle'$ state. There are two ways to do this: either a repump beam that drives $|1, 1\rangle$ to $|2, 2\rangle'$ down to $|2, 2\rangle$ can be used for a few μs before the image from the probe beam is acquired, or a microwave pulse can be used to transfer the atoms to $|2, 2\rangle$ without light. We found it to be very important that we use the latter approach, as the former method scatters several photons which have a probability of causing atoms to tunnel due to the heating, which in turn makes the cloud appear larger.

The size in the object plane corresponding to one pixel (referred to as the pixel size) can be estimated simply from the resolution and the magnification of the imaging system. However, a more accurate measurement will be needed to correctly count the number of atoms and the size of the cloud. For side-imaging paths, this can easily be done by measuring the vertical position during expansion. The cloud falls ballistically, and so the position in pixels can be fit to a quadratic polynomial and compared to $1/2 \, gt^2$. For vertical paths, as we used for all of our filling measurements with (relatively) high resolution, this technique does not work. The alternative we used was to look at the expansion of a superfluid from the lattice. The position of the interference peaks relative to the center could be fit to $2\hbar k \times t$, where t is the expansion time.

6.4.3 Tunneling During the Probe Time

Typically, in situ imaging is done with short probe times so that there is no atomic motion during the probe. This often requires a large intensity $I \approx I_{\text{sat}}$ simply to get enough light on the camera. We find empirically that probe times longer than $\approx 10 \, \mu s$ cause the cloud to appear larger. We can quantify this a bit more by considering the heating rate from the probe beam, and the thermal tunneling rate that it induces.

The heating rates are calculated using the photon scattering rates given by:

$$\Gamma_{\text{sc}} = \left(\frac{\Gamma}{2}\right) \frac{(I/I_{\text{sat}})}{1 + 4(\Delta/\Gamma)^2 + (I/I_{\text{sat}})}, \tag{6.6}$$

where Γ is the linewidth of the electronic excited state, and Δ is the detuning from resonance of the excited state. The heating rate can be calculated from this scattering rate via $\dot{T} = 2/3 \times (2 \times E_r)/k_B \times \Gamma_{\text{sc}}$, where E_r is the recoil energy [14]. For a probe beam of intensity $I = I_{\text{sat}}$ and detuning $\Delta = 0$, the scattering rate is $\Gamma_{\text{sc}} \sim 10^7 \, s^{-1}$, which corresponds to a heating rate of $\sim 2 \, \mu K/\mu s$ for Rb in the $|2, 2\rangle$ state ($\sim 4 \, \mu K/\mu s$ for K in $|9/2, -9/2\rangle$).

Since typical optical lattice depths are ~ 100s of μK, it seems very possible that atoms could undergo significant thermal hopping during the probe time. We can quantify this by approximating the lattice as a harmonic trap and defining a thermal hopping rate Γ_h [35]:

$$\Gamma_h \approx \Gamma_{\text{sc}} \left(\int_{V_{\text{latt}}}^{\infty} e^{-E/k_B T} dE \right) / (k_B T) = \Gamma_{\text{sc}} \sqrt{\pi} \, \text{erfc}(\sqrt{V_{\text{latt}}/k_B T}). \tag{6.7}$$

An upper bound to this hopping rate is twice the oscillation frequency, which is the frequency of hopping attempts.

Assuming that the lattice depth is $s_{\text{Rb}} = 50$, the trap oscillation frequency is $\omega_{\text{Rb}} \approx 20 \, \text{kHz}$. Further, assume that the initial temperature (kinetic energy) in the lattice is slightly higher than the temperature of the initial BEC, so $T_i = 200 \, \text{nK}$. The above equation can be used to estimate the thermal hopping rate for this case, and gives $\approx 10^{-6} \, s^{-1}$, while the coherent tunneling rate is $\approx 0.1 \, s^{-1}$. Therefore, at this temperature thermal hopping is insignificant, but it will drastically increase as the atoms are heated during the probe pulse, and the band index in the lattice starts

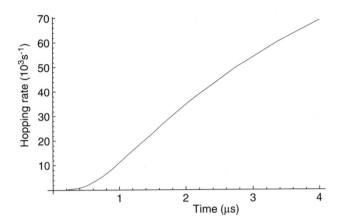

Fig. 6.8 The thermal hopping rate as a function of the probe time. The integral under the curve gives a hopping probability of ≈15% in 4 μs

to become large. The thermal hopping rate as a function of the probe duration is shown in Fig. 6.8, and it is calculated using the heating rate from the probe beam given above. The area under this curve gives a rough estimate of the tunneling probability during the probe time. For these parameters, this probability is ≈15% in 4 μs. This is sufficiently low to avoid significantly perturbing the size of the cloud, but it illustrates why a short probe pulse is important.

6.4.4   Molecule-Specific Systematics

There are also effects that arise from having two species and two atoms per site which are important to consider. Specifically, it is important that the presence of one species does not affect the counting of the other species. This could happen through photoassociation loss during imaging, or by inelastic spin changing collisions, to be discussed later in this chapter. Here, I will focus on the following point: once the molecules are dissociated into doublons (i.e., sites with one K and one Rb), we should be able to measure the same number and filling fraction of both species. If we cannot do this, then it will be difficult to trust any measurements of the filling fraction of molecular clouds. References [5, 22] focus on this point, and we were able to demonstrate that we do indeed measure the same filling fraction and atom number of both species. Moreover, we demonstrated that we can image the species in either order, which further strengthens our confidence in our filling measurements of K, Rb, and KRb.

6.5 Simulating the Rb Mott Insulators

In addition to the excellent understanding of how to probe atoms in situ with high fidelity, it is important to develop a theoretical understanding of what we expect the atomic distributions to be in our lattice. To do this, we return to the discussion in Chap. 4 of the superfluid to Mott insulator transition. The simulations we perform are at $T = 0$, which is appropriate for all of our data. However, finite temperature corrections were considered in the group of Ana Maria Rey [28], and are used later in the discussion of the doublon fraction.

6.5.1 Calculating the Atomic Distributions at $T = 0$

For a perfect, $T = 0$ Rb MI, we can calculate the distribution without tunneling, which is based on the analysis of Ref. [9]. This allows us to numerically find the relationship between the chemical potential μ_0 and the particle number N, and the local chemical potential is given by $\mu(i, j, k) = \mu_0 - V(i, j, k)$, where (i, j, k) are the site indices for (x, y, z) and V is the harmonic confinement. In the absence of tunneling, the occupancy n on site (i, j, k) must satisfy

$$(n - 1) < \frac{\mu(i, j, k)}{U} \leq n, \tag{6.8}$$

where U is again the on-site interaction between two atoms. The green staircase in Fig. 6.10a is exactly this calculation, given the uncertainty in the harmonic confinement, which is $\omega_r = 2\pi \times 38(2)\,\text{Hz}$ and the aspect ratio of the trap is $A = 6.4(1)$.

Figure 6.9a shows how the atoms are distributed in the lattice as a result of such a simulation. The number of atoms integrated through the vertical direction is shown as a function of the radial coordinates. Figure 6.9b shows the peak filling in the center of the distribution as a function of the atom number. Note that the staircase behavior is washed out because the distribution is convolved with the imaging resolution, to be discussed next. Figure 6.9c shows a cross section cut through the simulation in Fig. 6.9a, where the different colors show cross sections at varying distances from the center. A Gaussian fit to the center cross section is shown in Fig. 6.9d, and demonstrates what the distribution looks like when fit with a Gaussian function.

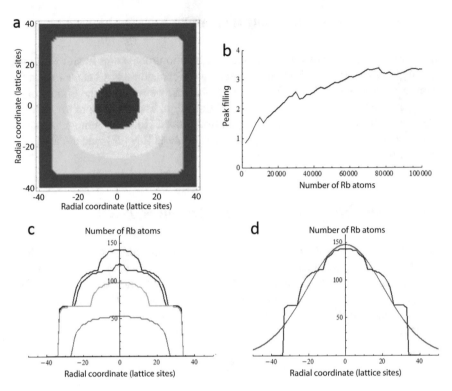

Fig. 6.9 Simulating the Mott insulator shells. (**a**) The number of atoms integrating through the vertical direction. The axes denote lattice sites, and the Mott shells are clearly apparent. (**b**) The peak filling in such simulations as a function of the total atom number. Excellent agreement with experimental data can be seen in later figures. (**c**) A cross section cut through the simulation in (**a**). The different colors show cross sections at varying distances from the center. (**d**) A Gaussian fit to the center cross section shows what the distribution looks like when convolved with the experimental imaging resolution

6.5.2 Convolving the Atomic Distributions with the Imaging Resolution

We compare these simulations to experimental data by convolving them with a Gaussian filter which has RMS width of 4.5(5) lattice sites, in accordance with our resolution. We also account for the pixelation due to the finite imaging magnification by mapping an area of 6×6 lattice sites onto one pixel. This gives us a convolved, pixelated 2D distribution that we then fit with a 2D TF surface to extract f_{Rb}. This is shown as the orange band in Fig. 6.10a. The width of the band accounts for uncertainties in the trap, the resolution, and the pixel size. We found it more rigorous to compare the pixelated and convolved theory to the data as opposed to trying to deconvolve the data and reverse-pixelate it. Obviously, a lot of information is missing, so this approach requires an assumption of a TF distribution anyway.

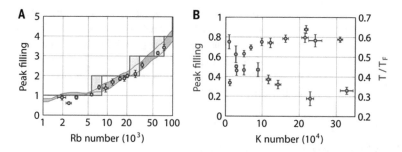

Fig. 6.10 The filling fractions vs. their respective atom numbers. (**a**) The peak filling (filling in the center of the atomic distribution) of Rb as a function of the Rb number. The green lines show the zero temperature theory for filling, and the orange band shows the same theory predictions under a finite imaging resolution realized in the experiment. (**b**) The peak filling in the lattice (blue circles, left vertical axis) and temperature of ^{40}K prior to loading the lattice (red circles, right vertical axis) as function of ^{40}K number. The temperature is normalized to the Fermi temperature of the ^{40}K gas in the optical harmonic trap. Reproduced from Ref. [22]

6.5.3 Calculating the OD Per Atom

In order to compare the calculations of the number of atoms per site with the measured ODs, it is important to understand the OD per atom. The OD per area is given by $\tau = n \times \sigma_0$. $n = N/A$ is the 2D area of the cloud integrated through the z-direction. To consider a single atom, the number is $N = 1$, and the area is the size of the atomic wavefunction on the lattice site. Based on the discussion in Chap. 4, for the typical value of $s = 25\ E_r$ and $a = \lambda/2 = 532$ nm, the wavefunction size $\gamma = 75$ nm in the ground band. However, during imaging there is significant heating at a rate $\approx 1\ \mu\text{K}/\mu\text{s}$, so the size will quickly get much bigger. The upper limit is the size of the lattice site $\lambda/2 = 532$ nm, so we can assume an atom size of $\gamma \approx 100$–200 nm. The absorption cross section for circular polarization is given by $\sigma_0 = 3\lambda^2/2\pi$, but we use linear polarization which reduces this cross section by a factor of two. Combining these factors together provides an estimate of $\tau \approx 0.5 - 1/\text{atom}$.

6.6 The Atomic Peak Filling Fractions

In Ref. [22], we first studied the filling of the atomic gases separately and determined that we need to work with a very small MI (fewer than 5000 Rb atoms) and a large Fermi gas (more than 10^5 ^{40}K atoms) to achieve fillings approaching one atom of each species per site for the given external confinement potential (see Fig. 6.10). For the Fermi gas, we reached the band insulator limit where the filling is saturated beyond 10^5 ^{40}K atoms where the Fermi gas is much larger than the Bose gas (see Fig. 6.5).

A band insulator at $T = 0$ is characterized as an incompressible many-body state of spin-polarized fermions in the lattice where every site has a fermion, and thus tunneling is suppressed by the Pauli exclusion principle [19]. While $T = 0$ is an appropriate approximation for Rb, the Fermi gas is relatively hot and the band insulator is thus riddled with thermal holes. Thus, the filling fraction saturates at only $\approx$80–90% even for the band insulator.

6.7   Measuring the Temperature of K

It is useful to understand what the temperature of the K gas is before loading into the lattice, as this will ultimately determine the filling fraction that we can reach. In general, it is difficult to accurately measure the temperature of a degenerate Fermi gas (DFG), because its profile in expansion deviates from a Gaussian at low temperatures. Therefore, a poly-log fitting function with a fugacity must be used [26]. Alternatively, the cloud in expansion can be truncated to remove atoms within $\approx$50% of the maximum OD, and then the wings can be fit to a Gaussian to obtain the temperature, since the deviation of a Fermi–Dirac distribution from a Gaussian is maximized at the center. Then, the number can be obtained by fitting the entire cloud, from which $T_F = \hbar\bar{\omega}/k_B \cdot (6N)^{1/3}$ can be calculated [26]. $\bar{\omega}$ is the geometric mean trap frequency.

While we used both approaches and found that they give consistent results, we preferred to use a third approach in which we use Rb as a thermometer of the K cloud. To do this, we hold the Rb BEC and the K DFG in the trap with finite interspecies interaction. Within 100s of ms, the BEC will melt, and equilibrate with the hotter K gas. This temperature can be measured very accurately, and we wait until the BEC fraction becomes <20%. This gives the T for K, and N for K can be measured by fitting the entire K gas, thus giving T/T_F.

Figure 6.10b shows the temperature of K as well as the peak filling as a function of the K number. The trend in T/T_F is actually very strongly anti-correlated with the trend in f_K. We understand this phenomenon by considering that when the trap is very weak at the end of evaporation, the Rb BEC fails to cool the DFG further. Nevertheless, $T_F \sim N^{1/3}$, and thus T/T_F continues to go down for increasing N_K until it saturates at $\approx$2 × 10^5 at a value of $T/T_F = 0.3$.

The entropy per particle in this Fermi gas can be calculated and compared with the entropy per particle that corresponds to 80–90% filling in the lattice. The latter is calculated using $S/N = -k_B/f \cdot [f\ln(f) + (1 - f)\ln(1 - f)]$ [22], and gives $0.6k_B$. Using the finite temperature calculations for the K and Rb gases loaded into the lattice in Ref. [28], we find that our initial T/T_F is in good agreement with the filling fraction in the lattice. This suggests that our loading procedure of ramping up the lattices with a smooth, s-shaped function in 200 ms is sufficiently adiabatic.

Further cooling of the DFG would require more efficient thermalization at the bottom of the trap, which is very challenging once the Rb BEC becomes quite pure.

An alternative which we are considering for the new experimental apparatus is to have K efficiently cool itself at the end of evaporation by transferring some of the atoms to $|9/2, -7/2\rangle$ so that s-wave collisions can be used for K to thermalize with itself, and then we can evaporate K as well as Rb by selectively removing the hottest K atoms using a combination of gravity and a magnetic field gradient. We have never tried this in the old apparatus, but we intend to return this idea in the new apparatus in an effort to reach >90% filling of K in the lattice.

6.8 The Role of Interspecies Interactions

In order to preserve the filling of the Rb MI in the presence of such a large ^{40}K cloud, it is imperative to turn the interspecies interactions off by loading the lattice at $a_{K\text{-}Rb} = 0$. This is shown in Fig. 6.11. The top shows the filling fraction of Rb in the lattice normalized to what it is without K as a function of the scattering length. The background scattering length is shown as the vertical dotted line, so that the data on the left of the line is above the resonance. The bottom shows the BEC fraction with K prior to loading the lattice also vs. the scattering length. The red band shows the BEC fraction of Rb alone.

Note the clear correlation between the two curves, which suggests that the interaction between K and Rb deteriorates the BEC and thus has deleterious effects on the Rb filling fraction. We attribute this heating of the BEC to the large heat load

Fig. 6.11 The effect of K on the BEC and filling fraction of the initial Rb. The top shows the filling fraction of Rb in the lattice with normalized to what it is without K as a function of the scattering length. The background scattering length is shown as the vertical dotted line, so that the data on the left of the line is above the resonance. The bottom shows the BEC fraction with K prior to loading the lattice also vs. the scattering length. The red band shows the BEC fraction of Rb alone. Note the clear correlation between the two curves. Reproduced from Ref. [22]

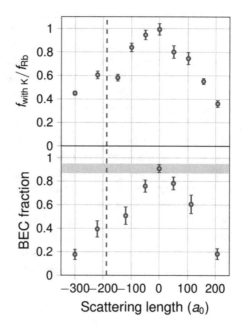

of the K Fermi gas which is at a relatively high temperature of $T = 0.3T_F$. Once Rb forms a BEC, it thermalizes with K very poorly, and the K number is ~50× the Rb number for this data. Thus, enabling interactions before loading the lattice melts the BEC and distorts the filling for a trivial reason. While we have studied the effects of K on the superfluid to Mott insulator transition of Rb as shown in Chap. 4, we do not have a definitive understanding of the role of interactions [2], even though they are expected to play a significant role in the lattice loading [11, 28]. Thus, we work at zero K–Rb scattering length.

However, the magneto-association proceeds by ramping the magnetic field from high to low values across the Feshbach resonance, while the location of $a_{K-Rb} = 0$ is below the resonance (see Fig. 6.12a). Hence, before we can perform magneto-association we need to first ramp the magnetic field from where $a_{K-Rb} = 0$ is located to above the resonance. To avoid populating higher bands when we do this, we first transferred ^{40}K to the $|F, m_F\rangle = |9/2, -7/2\rangle$ state to avoid the Feshbach resonance. This procedure is depicted in Fig. 6.12a. We then ramped the B field above the resonance, and transfer the ^{40}K atoms back to the $|9/2, -9/2\rangle$ state. Finally, we converted the K–Rb pairs into Feshbach molecules on each site and we studied the fraction of Rb atoms that were converted to molecules (Rb is the minority species).

6.9   Maximizing the Doublon Fraction

In the limit of a small Rb atom number, we converted more than 50% of the Rb to FbM (Fig. 6.12b). We then produced ground-state molecules and removed all unpaired atoms from the lattice. The final number of ground-state molecules can be determined by reversing the transfer process and then performing the normal absorption imaging on atoms. Typical in situ images are shown in Fig. 6.12c. In the case of high conversion, we achieved molecule fillings of at least 25%, which is significantly higher than our previous work [16, 38]. With some quick experimental improvements, we should be able to increase the filling even further.

This filling fraction is at the percolation threshold, where every molecule would be connected to every other molecule in the entire lattice [22], and it marks the first time that the polar molecules have entered the quantum degenerate regime, with an entropy per particle of $2.2k_B$ (k_B is Boltzmann's constant). Studying the spin dynamics described in the previous section in this higher-filled lattice will be the subject of future work. We note that a similar procedure has recently been pursued in the RbCs experiment in Innsbruck, where an Rb superfluid was moved to overlap a Cs Mott insulator before increasing the lattice depth to produce a dual Mott insulator. This technique has led to a gas of RbCs Feshbach molecules with 30% filling [27] but has not yet been combined with STIRAP to make ground-state molecules.

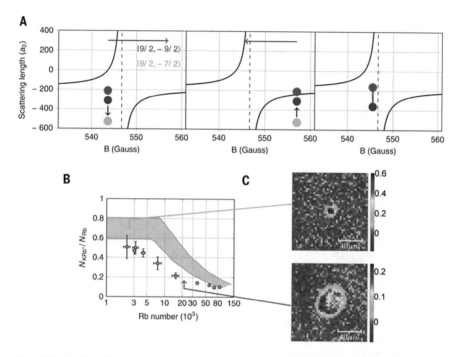

Fig. 6.12 (**a**) The K–Rb scattering length as a function of the magnetic field. Initially, the K atoms are prepared in the resonant state $|9/2, -9/2\rangle$. We spin flip the K atoms to the non-resonant $|9/2, -7/2\rangle$ state, before we sweep the magnetic field to above the Feshbach resonance, at which time we drive the K atoms back to the $|9/2, -9/2\rangle$ state. We can then sweep the field to below the resonance to make Feshbach molecules. (**b**) The number of Feshbach molecules produced normalized to the initial Rb number versus the number of Rb atoms. Theory band corresponds to calculations based on the zero temperature curve in Fig. 6.10a. (**c**) Images of clouds of ground-state molecules in the optical lattice for high and low filling fraction. Reproduced from Ref. [22]

6.10   An Alternative Measurement of the Doublon Fraction

The previous studies suggest that $\approx$50–60% of the Rb atoms are on a doublon site, but the actual conversion should only be limited by the filling of K, which is 80–90%. Therefore, it is unclear why the FbM filling, or equivalently the conversion from Rb to FbM, is low. To address this question, we have developed an alternative method for measuring the doublon fraction. Instead of associating doublons into FbM, we can turn on strong inelastic loss which causes doublons to decay in a few ms. This is done by flipping the Rb spin from $|1, 1\rangle$ to $|2, 2\rangle$, which is $h \times 8.1$ GHz higher in energy at 550 G. Since Rb and K ($|9/2, -9/2\rangle$) are stretched on opposite sides, angular momentum conservation does not help, and energy conservation is the only law which makes the Feshbach-resonant mixture stable, since $|1, 1\rangle$ and $|9/2, -9/2\rangle$ are the lowest energy states. Once $h \times 8.1$ GHz of internal energy is added by flipping Rb to $|2, 2\rangle$, the system decays very quickly, with $\beta = 6 \times 10^{-12}$ cm^3 s^{-1} [5].

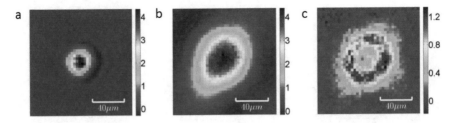

Fig. 6.13 Inelastic collisions between K and Rb cause loss of doublons. (**a**) An in situ image of a Mott insulator of $\sim 2 \times 10^4$ Rb atoms in an optical lattice. (**b**) An in situ image of a band insulator of K in an optical lattice. (**c**) An image of K after the inelastic loss of doublons. The hole in the middle represents the location of the Rb atoms and hence the doublons. Reproduced from Ref. [22]

By looking at the number of atoms lost from this inelastic decay, we can learn the number of doublons. Figure 6.13 shows how these measurements are made. Figure 6.13a shows a Mott insulator of Rb in the lattice with $N_{Rb} = 2$–3×10^4, and Fig. 6.13b shows a band insulator of K in the lattice with $N_K \approx 2 \times 10^5$. Once the inelastic loss is initiated, all the doublons are lost and a hole appears in the center of the K cloud which is the same size as the Rb cloud, as shown in Fig. 6.13c. This measurement also serves to verify that the two clouds are well overlapped, an obvious but non-trivial requirement.

Such images contain a wealth of information, most of which can be gleaned by counting the Rb number before and after the inelastic loss as a function of time. To describe this idea, let us first assume the filling of Rb is $f_{Rb} \leq 1$. Then, the fraction of Rb in a doublon will be equal to the filling fraction of K. In Fig. 6.14a, we consider this case, and once we turn on the inelastic loss we see that $\approx 80\%$ of the Rb is lost within $\tau = 2$ ms ($V_{latt} = 25\ E_r^{Rb}$). This suggests that 80% of the Rb were doublons and 20% were alone. Therefore, this measurement corroborates the measurement of f_K discussed above by in situ imaging of the K.

The residual decay on the longer timescale is also very interesting though we did not study it thoroughly. Since the K cloud is larger than the Rb cloud, the Rb atoms that remain after the doublon decay are alone inside a shell of K atoms. For $V_{latt} = 25\ E_r^{Rb}$, the lattice depth is only $10\ E_r^K$, and so the K tunneling rate is $J_K^0 = 90$ Hz. Therefore, the K tunnels around quickly and can tunnel onto sites with an immobile Rb, which are then lost quickly before the K can tunnel again. While this setting sounds as though it could give rise to the continuous quantum Zeno mechanism which would prevent the K from tunneling onto sites occupied with Rb (similarly to what we observed with a spin mixture of molecules as discussed in Chap. 4), the on-site inelastic loss rate is too low for the Zeno regime.

Let us now consider the case when $f_{Rb} > 1$, where sites with two Rb atoms and one K atom will be common. The loss on these sites will still be dominated by pairwise loss of a K and an Rb. Indeed, three-body loss of Rb does not become significant until we reach four Rb per site [3]. Therefore, we expect the inelastic loss of Rb to be less than f_K. This is shown in Fig. 6.14b, where the loss is $\approx 50\%$. This

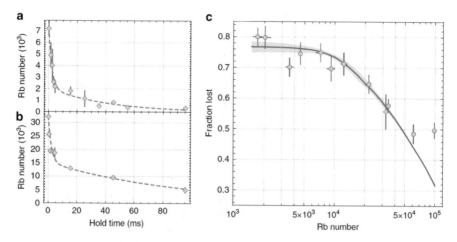

Fig. 6.14 An alternative method for determining the atomic filling fractions. (**a**) and (**b**) show the loss of Rb from inelastic collision with K as a function of time for low and intermediate number, respectively. (**c**) shows the fraction lost in ∼2 ms as a function of the initial Rb number. Reproduced from Ref. [5]

trend can be measured for all Rb numbers, and is shown in Fig. 6.14c. Notice that the data indicates a corner where the fraction lost drops below f_K, and this corner exactly matches the onset of significant double occupancy of Rb.

This technique tells us both the filling fraction of K and the number of Rb which corresponds to the onset of double occupancies, and both of which are in excellent agreement with the in situ measurements discussed above. The curves in Fig. 6.14a, b are dashed because they are simply double exponential fits. The curve in Fig. 6.14c is a finite temperature simulation of the Rb distribution. The width is due to the uncertainty in the harmonic trapping frequency and the uncertainty in the trap aspect ratio, and the fitting parameter is f_K. Note that the data starts to deviate from theory when $f_{Rb} \approx 4$ at $N_{Rb} \approx 10^5$, because three-body loss of Rb becomes significant and serves to overestimate the fraction loss.

This doublon decay technique confirms that indeed 80% of the Rb atoms are in doublons, and so when $f_{Rb} = 1$, the doublon filling fraction should be 0.8. Yet, we only convert ≈50% of the Rb to Feshbach molecules. This discrepancy is very important to understand, and could originate from two sources. First, since the tunneling rate of K is high and there are an enormous number of unpaired K atoms, it is possible that some K atoms are tunneling onto sites with FbM and then quickly undergoing inelastic decay. The other possibility is that the efficiency with which a doublon can be converted to an FbM is less than unity. We believe the latter effect to play a more significant role for these particular experiments, and so we must devise a way to measure the magneto-association efficiency of doublons in the optical lattice (however, the former problem could be solved using a colder K cloud which requires less excess K atoms to enter the band insulating regime).

6.11   Magneto-Association Efficiency on Lattice Sites

Even if every site has one ^{40}K and one Rb (i.e., a doublon), a highly filled lattice of molecules can only be achieved if the conversion of doublons to Feshbach molecules is very efficient. We already know that the STIRAP efficiency of converting Feshbach molecules to the ground state is about 90%, so we set out to develop another experimental technique to allow us to investigate this Feshbach molecule conversion efficiency.

Once the KRb molecules are produced in the ground state, unpaired atoms can be removed with resonant light. In principle, this atom-removal process does not affect the molecules. However, the fact that a large number of ^{40}K atoms is required and that KRb molecules can chemically react with ^{40}K if they encounter each other at short range does lead to a loss of a small fraction of ground-state molecules. Nevertheless, after the free atoms are removed, the 3D optical lattice now has sites that are either empty or contain one ground-state molecule. The STIRAP process can then be reversed, which leads to sites that are either empty or contain a Feshbach molecule. Then, the field is swept back above the Feshbach resonance to dissociate the molecules, leading to a scenario where lattice sites are either empty or contain a Bose–Fermi atomic pair, referred to as a doublon. This process is outlined in Fig. 6.15, and the preparation of this clean initial condition can be used to study the conversion efficiency of doublons into Feshbach molecules.

We have identified effects from a narrow, higher-order d-wave Feshbach resonance near the broad s-wave resonance we use for molecule association (shown in Fig. 6.16a). In the doublon preparation step, the speed of the dissociation sweep from below to above the resonance is extremely important. Figure 6.16b shows the consequences of the d-wave resonance. The dissociation starts in the green channel at (1). Then, the rate of the magnetic field ramp determines whether the molecule converts to doublons in (2) or (3). The case of (2) is when the d-wave resonance is crossed adiabatically, and the FbM are connected to a doublon in the ground band of the optical lattice. However, when the ramp is too slow and the d-wave resonance is crossed adiabatically, the FbMs connect to (3) where there is a motional excitation in the doublon, and a weighted superposition of K and Rb is in the first excited band in the optical lattice.

Then, the Feshbach association measurement is conducted by sweeping from above the resonance back to the other side of the s-wave resonance, and this is done very quickly ($\approx$10–20 G/ms). For ground-band doublons in (2), the fast sweep is diabatic with respect to the d-wave resonance, and they are converted to s-wave FbM. For the excited-band doublons in (3), the fast ramp rate back below the resonance is also diabatic with respect to the d-wave resonance and the doublons are not connected to any FbM, but rather a doublon in the ground band.

The results of this measurement, where the ramp rate from below to above is varied and the ramp rate from above back to below is very fast, are shown in Fig. 6.16c.

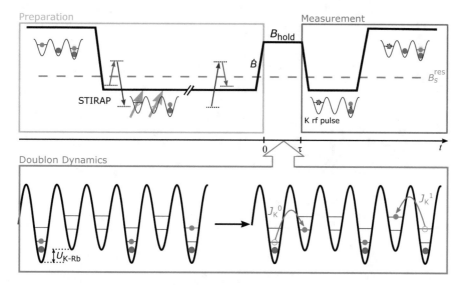

Fig. 6.15 The experimental setup for producing K–Rb doublons in an optical lattice. The first step is the preparation, where a pure sample of molecules in the optical lattice is turned into sites with one K and one Rb using the STIRAP sequence. We then let these doublons evolve in the lattice for a variable time at a variable scattering length. During the evolution, both K and Rb can tunnel (potentially together), but for short evolution times only K will tunnel since it feels a weaker lattice due to its smaller mass. After the evolution period, we associate doublons that have not fallen apart back into Feshbach molecules, which can be spectroscopically differentiated in the measurement from unpaired K atoms that have separated from their doublon Rb partner. Reproduced from Ref. [5]

This measurement is fitted with a Landau–Zener description $P = e^{-A/|\dot{B}|}$, where P is the probability of creating an s-wave FbM, $\dot{B}$ is the rate of the magnetic field ramp, and

$$A = \frac{4\sqrt{3}\omega_{\mathrm{HO}}|a_{\mathrm{bg}}\Delta_d|}{L_{\mathrm{HO}}}, \tag{6.9}$$

where ω_{HO} is the harmonic oscillator trapping frequency on a lattice site, $L_{\mathrm{HO}} = \sqrt{\hbar/(\mu\omega_{\mathrm{HO}})}$ is the harmonic oscillator length with the doublon reduced mass μ, $a_{\mathrm{bg}} = -187(5)a_0$ is the background scattering length of the s-wave resonance, and Δ_d is the width of the d-wave resonance [5]. This data provides the most accurate measurement of the width of this resonance, $|\Delta_d| = 9.3(7)$ mG [5]. In Fig. 6.16d, we ramp slowly to various fields to identify to position of the resonance as 547.47(1) G. This is the most accurate determination of its position.

Historically, the ramp rate we have always used in the lattice is 3.4 G/ms (0.34 mT/ms), and thus this narrow resonance was causing our Feshbach molecule creation efficiency to be 70% since we first produced molecules in an optical lattice [4]. When we actually make FbM, we sweep from above to below, and so

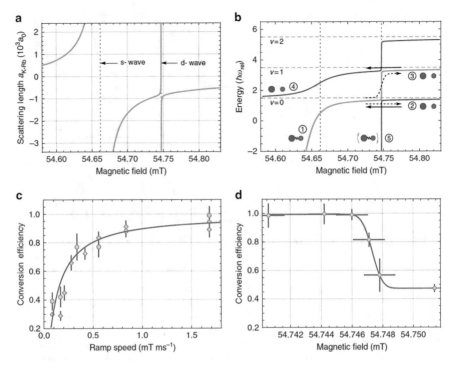

Fig. 6.16 Characterizing the d-wave resonance. (**a**) shows the scattering length as a function of magnetic field, for both the s-wave and d-wave resonance. (**b**) shows the adiabatic mapping of the bound and unbound states below the resonances to their corresponding harmonic oscillator states in the lattice above the resonance. (**c**) shows the probability of crossing the *d*-wave resonance adiabatically (i.e., not making a d-wave molecule) as a function of the ramp speed. (**d**) shows the same efficiency as a function of the field in order to identify its center. Reproduced from Ref. [5]

$\approx$30% of the FbM molecules produced in our experiment have always been *d*-wave FbM. These FbM will stay dark during the subsequent STIRAP sequence because their initial rotational angular momentum is $L = 2\hbar$. As a result of this, of the doublons which account for 80% of the Rb, only 50–60% are converted to s-wave FbM, which is consistent with our measurements.

By simply ramping across the resonance sufficiently fast so that we avoid making *d*-wave molecules adiabatically, we are able to solve this problem. With this mechanism understood, we anticipate 100% conversion efficiency for all future experiments. This improvement alone should allow us to reach GSM fillings of $f_{GSM} > 0.5$. I want to emphasize that since the *d*-wave resonance is only 9 mG wide, it would play no role in molecule formation in a harmonic dipole trap. However, since the on-site density in an optical lattice is so high, the timescales for keeping the sweep diabatic with respect to this resonance are much faster. We believe this to be the only investigation to date of narrow heteronuclear Feshbach

resonances in an optical lattice. Note, however, that similar resonances have been
explored in Cs_2, and navigation through the state manifold was carried out in
Ref. [7].

6.12   Atomic Tunneling: Doublon Dynamics

We have also investigated tunneling dynamics and interaction effects between the
bosonic and fermionic atoms composing the doublons. Tunneling dynamics of
doublons in the lattice can also affect molecule production. In the quantum synthesis
approach described above, achieving a high lattice filling for molecules requires not
only the preparation of a large fraction of lattice sites that have doublons, but also
that these doublons are not lost due to tunneling and/or collisions prior to conversion
to molecules. At $\lambda_{latt} = 1064$ nm, K feels a lattice depth that, in units of recoil
energy, is 2.6 times weaker than for Rb due to differences in atomic mass and
polarizability. Consequently, K tunnels faster than Rb.

Figure 6.17 illustrates doublon dynamics due to the interplay between tunneling
and interactions, which we control by varying the lattice depth and interspecies
scattering length $a_{K\text{-Rb}}$ through B_{hold}. We note that for $a_{K\text{-Rb}} > -850a_0$, the B
sweep crosses the d-wave resonance with a $\dot{B}$ that varies from 5 to 19 G/ms, and
so the data in Fig. 6.17 has been multiplied by a factor that increases the doublon
fraction to account for the finite $\dot{B}$ when crossing the d-wave resonance.

Figure 6.17a shows the effect of the lattice depth for $\tau = 1$ ms at three
different values of B_{hold}, corresponding to different values of $a_{K\text{-Rb}}$. This timescale
is relevant for both molecule production and K tunneling dynamics. We observe
that the remaining doublon fraction is highly sensitive to the lattice depth for weak

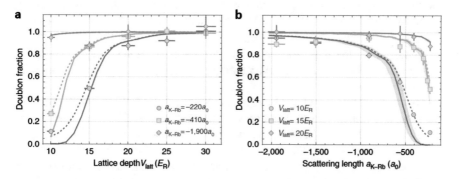

Fig. 6.17 K tunneling dynamics. (**a**) The survival probability for doublons in a lattice for 1 ms
as a function of lattice depth, shown for several intra-doublon interaction strengths. (**b**) The same
survival probability as a function of scattering length for several lattice depths. The solid lines
represent single doublon calculations, while the dotted lines show a lattice with $\sim$15% doublon
filling, where the doublons can interact and enhance their survival. Reproduced from Ref. [5]

interspecies interactions, e.g., $a_{K\text{-}Rb} = -220a_0$, with a lower doublon fraction for shallower lattices that exhibit higher tunneling rates. For stronger interactions, the dependence on lattice depth becomes less significant and almost disappears in the strongly interacting regime, e.g., $a_{K\text{-}Rb} = -1900a_0$. Similar behavior is observed if we fix the lattice depth but vary the interspecies interaction, as shown in Fig. 6.17b.

The data in Fig. 6.17 clearly shows evidence of decay of doublons due to tunneling that is affected by both the lattice depth and interspecies interactions. We can model doublon dynamics with the following Hamiltonian [5]:

$$H = -J_{Rb}^0 \sum_{\langle i,j \rangle} a_i^\dagger a_j - \sum_{\eta,\langle i,j \rangle} J_K^\eta c_{i,\eta}^\dagger c_{j,\eta} + \sum_{i,\eta} U_{KRb}^\eta n_{Rb,i}^0 n_{K,i}^\eta$$

$$+ \frac{U_{RbRb}^0}{2} \sum_i n_{Rb,i}^0 (n_{Rb,i}^0 - 1), \tag{6.10}$$

where $\eta = 0$ (1) denote the ground (first) excited band in the lattice. The first and second terms are the kinetic energy of the K and Rb atoms, respectively. Here, a_i $\left(a_i^\dagger\right)$ is the bosonic annihilation (creation) operator for an Rb atom at the lattice site i in the ground band, and c_i $\left(c_i^\dagger\right)$ is the fermionic annihilation (creation) operator for a K atom at lattice site i and band η. We use $\langle i, j \rangle$ to indicate nearest-neighbor hopping between sites i and j with matrix element J_α^η, where $\alpha = K$ or Rb. We assume $\eta = 0$ for Rb throughout, and only K in the excited band in considered. The third term describes the interspecies on-site interactions with matrix element U_{KRb}^η. The last term is the on-site intra-species interaction between ground-band Rb atoms with strength U_{RbRb}^0, with $n_{Rb,i}^0$ as the occupation of site i.

The solid lines in Fig. 6.17 show the calculations based on this Hamiltonian, where we have neglected Rb tunneling by setting $J_{Rb}^0 = 0$. We start with a single doublon, evolve the K for a hold time τ, and then extract the doublon fraction from the probability that the K atom remains on the same site as the Rb atom. In this treatment, we ignore the role of the magnetic-field sweeps. Calculations for a single doublon (solid lines), where the initial decay scales as $1 - 12(J_K^0/U_{KRb}^0)^2$ [5], agree well with the data, except at doublon fractions below ~30%, where the disagreement arises from the finite probability in the experiment that a K atom finds a different Rb partner.

The doublon fraction where this starts to matter depends on the doublon filling fraction, and could present another way to measure the FbM or GSM filling fractions. Simulating a Gaussian distribution of doublons with 10% peak filling, which is appropriate for this data, accounts for this effect. This data is shown as the dashed lines, and the good agreement of these calculations with the data shows that tunneling of K, which is suppressed in deeper lattices, is the dominant mechanism for the reduction of the doublon fraction at short (~1 ms) times. The on-site interaction with Rb suppresses the K tunneling when the interaction energy becomes larger than the width of the K Bloch band [5, 17].

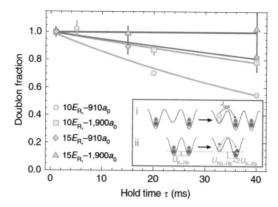

Fig. 6.18 K and Rb tunneling dynamics. The survival probability for doublons being held in the lattice for a variable time. Several combinations of lattice depth and interaction strength are shown, and the date is fit to simple exponentials. Note that the doublon fraction has been renormalized to the corresponding number in Fig. 6.17 to highlight the additional loss upon waiting longer times, which is due to Rb. The inset shows processes by which this additional loss can occur, and demonstrate a qualitative agreement with our data. Reproduced from Ref. [5]

Moreover, if sufficient tunneling time is allowed, we can even observe effects of doublon–doublon interactions. In Fig. 6.18, we present data taken for τ up to 40 ms in order to look for the effects of Rb tunneling. Measurements of the remaining doublon fraction are shown for two lattice depths ($10E_r$ and $15E_r$) and two values of $a_{\text{K-Rb}}$ ($-910a_0$ and $-1900a_0$). The doublon fraction has been normalized by the measured value for $\tau = 1$ ms in order to remove the effect of the shorter-time dynamics discussed above. Similar to the shorter-time dynamics, we see a reduction in the doublon fraction at long hold times which is suppressed by a deeper lattice and stronger interspecies interactions.

Modeling these dynamics is theoretically challenging, and the lines in Fig. 6.18 are exponential fits that are intended only as guides to the eye. Compared to doublons composed of identical bosons [37] or fermions in two spin states [33], the heteronuclear system has the additional complexities of two particle masses, two tunneling rates, and two relevant interaction energies. For example, for large $a_{\text{K-Rb}}$ the interspecies interactions will strongly suppress Rb tunneling from a doublon to a neighboring empty site. Similarly, tunneling of a doublon to an empty site is a slow second-order process at the rate $J_{\text{pair}} = 2J_{\text{Rb}}^0 J_{\text{K}}^0 / U_{\text{KRb}}^0$ due to the energy gap of U_{KRb}^0, as is shown in the inset.

However, Rb tunneling between two neighboring doublons, which creates a triplon (Rb–Rb–K) on one site and lone K on the other, may occur on a faster timescale due to a much smaller energy gap U_{RbRb}^0, which is smaller than the K tunneling bandwidth. While the theoretical description is complicated, we observe that the time scale of the doublon decay roughly matches $1/(2\pi J_{\text{pair}})$. Therefore, we believe that the loss of doublons on longer timescales comes from neighboring doublons, and the pair-hopping of next-nearest-neighbor doublons is the limiting step at long times when nearest-neighbor doublons have already interacted. This type of system could be very interesting for the study of quasi-crystallization and many-body localization in an optical lattice [5, 12, 30, 31].

6.13 The Role of Higher Bands in the Lattice

When studying doublon dynamics measured for two different initial atom conditions, we find direct evidence for excited-band molecules. We compare results for our usual molecule preparation using atomic insulators to a case where we start with a hotter initial atom gas mixture at a temperature above T_c or T_F (i.e., thermal gases). Using the band-mapping technique discussed in Chap. 4, we measure the initial population of K in the ground band and the first excited band, as shown in Fig. 6.19b. For this data, we found that 11(2)% of the K atoms occupy the first excited band for the colder initial gas (**i**), and 31(6)% of the K atoms occupy the first excited band for the significantly hotter initial gas (**ii**).

When looking at the doublon dynamics for these two cases (Fig. 6.19a), we observe a lower doublon fraction for the hotter initial gas for $V_{latt} \leq 25\ E_r^{Rb}$. These data are taken for 16.8 G/ms sweeps, $\tau = 1$ ms, and $a_{K\text{-}Rb} = -220a_0$. The lower doublon fraction can be explained by excited-band K atoms, which have a high tunneling rate (J_K^0/h and J_K^1/h are 89.3 and 1110 Hz, respectively,

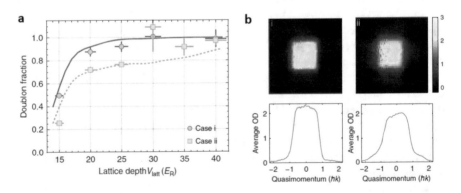

Fig. 6.19 The role of higher bands. (**a**) The doublon survival probability for a 1-ms hold vs. lattice depth with $a_{K\text{-}Rb} = -220\ a_0$. The Case **i** (red) data has a very low fraction of K atoms in excited bands prior to making molecules in the lattice. (**b**) The corresponding band mapping of the initial K, where a larger excited-band fraction is shown in Case **ii** (green). The green data in (**a**) is clearly lower, demonstrating a strong correlation between initial K in higher bands, molecules in higher bands, and dissociated molecules in higher bands leading to K in higher bands. The green dotted curve is the same as the red, except that the tunneling rate of ~20% (extracted from the data in (**b**) Case **ii**) of the doublons is the higher-band value instead of the ground-band value. Reproduced from Ref. [5]

for $V_{latt} = 25\ E_r^{Rb}$). The presence of excited-band K atoms suggests that the B sweeps for magneto-association (and dissociation) couple excited-band K atoms (plus a ground-band Rb atom) to excited-band Feshbach molecules. Moreover, the data suggests that the conversion efficiency for the excited-band Feshbach molecules is still high for $V_{latt} \le 25\ E_r^{Rb}$. This can be seen by the observed difference (roughly 20%) in the doublon fraction, which matches the difference in excited-band fraction of K between the two cases (as shown in Fig. 6.19a).

Since we prepare the doublons directly from the dissociated GSM, these results further indicate that a polar molecule sample prepared from a finite-temperature gas can contain a small fraction of molecules in an excited motional state in the lattice. We also observe that an Rb excited-band population of 31(5)% after loading the thermal gas in the lattice; however, even in the excited band, the off-resonant Rb tunneling is slow compared to the 1-ms timescale of these measurements. The green dashed curve in Fig. 6.19a shows the theoretical results for a K excited-band fraction of 24%. For comparison, the red solid curve, which is the same as the red curve in Fig. 6.17a, includes no excited-band population.

The estimated excited-band fraction ignores the effects of harmonic confinement on tunneling, which are more significant for the hotter initial atom gas, where the resulting molecular cloud is also larger. For the hotter initial atom gas, the green dashed curve overlaps the data at the shallower lattice depths but deviates from the measured doublon fraction at larger lattice depths (the excited-band fraction of the initial K gas is independent of lattice depth). This may be expected since in the limit of a very deep lattice and a fully adiabatic magneto-association sweep, one expects that only the heavier atom (Rb) in excited bands (plus a ground-band K atom) will couple to excited-band Feshbach molecules.

Further evidence for the correlation between excited-band atoms in the initial K gas and excited-band molecules created from it is shown in Fig. 6.20, where the measured KRb excited-band fraction is plotted versus the excited-band fraction of the K gases from which it was created for several temperatures of the initial atomic gases. Sample images of the band-mapped K and KRb clouds are shown for very low and very high initial temperatures. This trend corroborates all the indirect measurements based on doublon dynamics. However, the slope of the fitting line cannot be explained by the above reasoning based on the harmonic oscillator levels for excited K and Rb atoms in the lattice. Nevertheless, it is clear that the temperature of the initial atomic gases is highly correlated with excited-band fraction of the GSM and FbM they create.

It is also possible that even if all the FbM are in the ground band, the two-photon STIRAP sequence couples them to a GSM in an excited band. Typical STIRAP linewidths are $\sim$200 kHz, which is significantly larger than the band spacing of $\sim$20 kHz, and so the band structure is fully unresolved. To understand this possibility, we must consider the Lamb-Dicke factors for the lattice depth typically used to make molecules, which is $V_{latt} = 25\ E_r^{Rb} = 40\ E_r^{KRb}$. We define the wave vectors of the two Raman beams as $\vec{k}_u$ and $\vec{k}_d$ for the up leg and the down leg, respectively. The momentum transfer scales as $\sim e^{i(\vec{k}_u - \vec{k}_d) \cdot \vec{r}}$.

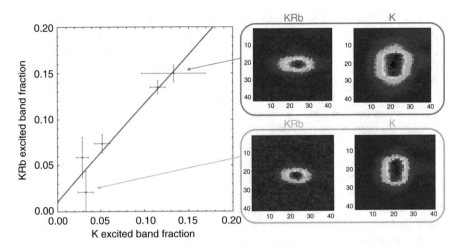

Fig. 6.20 Correlation between K and KRb in higher bands. The fraction of K atoms in higher bands initially loaded into the lattice is correlated with the number of molecules in higher bands in the lattice that were synthesized from these atoms. Sample band-mapping images are shown for K and KRb for both low and high excited-band fractions

For excited bands in the direction $\vec{\xi}$, this momentum transfer is $\vec{k}_\xi = (\vec{k}_u - \vec{k}_d) \cdot \vec{\xi}$. Moreover, the lattice spacing is $\lambda/2 = a = 532\,\text{nm}$. For the parameters typically used for STIRAP, we find that our experiment is well within the Lamb-Dicke regime $k_\xi a/s_{\text{KRb}}^{1/4} \ll 1$, where $s_{\text{KRb}} = V_{\text{latt}}^{\text{KRb}}/E_r^{\text{KRb}}$ [28]. Therefore, the total population in the first excited band due to STIRAP is ~1% [28]. This is for beams co-propagating at a 45° angle with respect to the x and y lattice axes and a 90° angle with respect to the z lattice axis. In the new apparatus (discussed in Chap. 7), the Raman beam will be co-propagating with a 0° angle with respect to the z lattice axis and 90° with respect to the x and y lattice axes. Nevertheless, this probability is still be negligibly small.

6.14   Photoassociation of Doublons in the Lattice

Another question we addressed with doublons is whether the FbM step could possibly be circumvented by efficiently photoassociating doublons to a bound state in the electronically excited potential, which could be coupled to the ground state. If it were possible to do this with high fidelity, we could even imagine doing STIRAP straight from a doublon to a GSM. To address this question, we study the effects of the up leg Raman beam on doublons in the lattice. Figure 6.21a shows the normal up leg transfer from FbM to the excited intermediate state of the STIRAP transition, which is dark to atomic detection. The pulse shown is ≈10 μs, which is consistent with the Rabi frequency of that transition and the transition dipole matrix element discussed in Chap. 2.

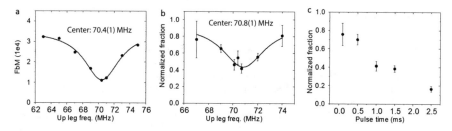

Fig. 6.21 Photoassociation of doublons in the lattice. (**a**) The lineshape of the up leg pulse on Feshbach molecules, taking them to a dark state which is not detected. (**b**) The up leg pulse on doublons (not associated to Feshbach molecules). The pulse times are very different between the two, reflecting the very different Franck–Condon Factors. Moreover, the frequency has shifted by 400 kHz, which matches the binding energy of the Feshbach molecules at this field. (**c**) The doublon photoassociation loss as a function of the up leg pulse time. Note that the time in (**a**) is ∼µs. The loss timescale here is incredibly long, suggesting that the Rabi frequency is incredibly small. Also note that the loss timescale matches the K tunneling rate in the lattice

Figure 6.21b shows the same curve except with doublons instead of FbM. While the width is similar, the center frequency has shifted by ≈400 kHz, which is consistent with the binding energy of the FbM at the field that we typically use. The pulse time in Fig. 6.21b is $\tau \approx 1$ ms, which indicates that the Rabi frequency is much lower. Figure 6.21c shows the fraction of doublons lost as a function of time, and it shows that the characteristic pulse time is $\tau \approx 1$ ms. However, this timescale is approaching the tunneling rate of K in the lattice, and so it is difficult to gain a quantitative understanding about the Rabi frequency of the photoassociation pulse.

Nevertheless, it is perhaps not surprising that the Rabi frequency would be so low since the Franck–Condon overlap of the doublon center-of-mass wavefunction and the intermediate STIRAP state is very poor. The molecular state of the FbM is very highly excited in the vibrational degree of freedom, so the wavefunction that describes the internuclear separation has many nodes. Conversely, if both K and Rb are in the ground band of the lattice, their motional wavefunctions have no nodes. Even though doublons adiabatically connect to FbM, their wavefunctions are quite different in size and shape.

Although the strong confinement of the lattice alters the properties of the Feshbach resonance and can lead to the so-called confinement induced resonances [15], the Franck–Condon overlap between such doublons and bound molecular states in the excited molecular potential is very small. Therefore, direct photoassociation as a first step of a Raman transition to GSM is not practical. It is worth pointing out, however, that the situation may be much more favorable in microtraps, called tweezers, rather than an optical lattice because typical oscillation frequencies are ∼MHz rather than ∼10s of kHz. This question is now being explored at Harvard [18, 20] and is quickly attracting more interest.

References

1. K. Aikawa, A. Frisch, M. Mark, S. Baier, R. Grimm, F. Ferlaino, Reaching Fermi degeneracy via universal dipolar scattering. Phys. Rev. Lett. **112**, 010404 (2014)
2. Th. Best, S. Will, U. Schneider, L. Hackermüller, D. van Oosten, I. Bloch, D.-S. Lühmann, Role of interactions in ^{87}Rb–^{40}K Bose-Fermi mixtures in a 3D optical lattice. Phys. Rev. Lett. **102**, 030408 (2009)
3. G.K. Campbell, J. Mun, M. Boyd, P. Medley, A.E. Leanhardt, L.G. Marcassa, D.E. Pritchard, W. Ketterle, Imaging the mott insulator shells by using atomic clock shifts. Science **313**(5787), 649–652 (2006)
4. A. Chotia, B. Neyenhuis, S.A. Moses, B. Yan, J.P. Covey, M. Foss-Feig, A.M. Rey, D.S. Jin, J. Ye, Long-lived dipolar molecules and Feshbach molecules in a 3D optical lattice. Phys. Rev. Lett. **108**, 080405 (2012)
5. J.P. Covey, S.A. Moses, M. Garttner, A. Safavi-Naini, M.T. Miecnkowski, Z. Fu, J. Schachenmayer, P.S. Julienne, A.M. Rey, D.S. Jin, J. Ye, Doublon dynamics and polar molecule production in an optical lattice. Nat. Commun. **7**, 11279 (2016)
6. B. Damski, L. Santos, E. Tiemann, M. Lewenstein, S. Kotochigova, P. Julienne, P. Zoller, Creation of a dipolar superfluid in optical lattices. Phys. Rev. Lett. **90**, 110401 (2003)
7. J.G. Danzl, M.J. Mark, E. Haller, M. Gustavsson, R. Hart, J. Aldegunde, J. M. Hutson, H.-C. Nägerl, An ultracold high-density sample of rovibronic ground-state molecules in an optical lattice. Nat. Phys. **6**, 265–270 (2010)
8. M.H.G. de Miranda, A. Chotia, B. Neyenhuis, D. Wang, G. Quéméner, S. Ospelkaus, J.L. Bohn, J. Ye, D.S. Jin, Controlling the quantum stereodynamics of ultracold bimolecular reactions. Nat. Phys. **7**(6), 502–507 (2011)
9. B. DeMarco, C. Lannert, S. Vishveshwara, T.-C. Wei, Structure and stability of mott-insulator shells of bosons trapped in an optical lattice. Phys. Rev. A **71**, 063601 (2005)
10. P. Dyke, K. Fenech, T. Peppler, M.G. Lingham, S. Hoinka, W. Zhang, S.-G. Peng, B. Mulkerin, H. Hu, X.-J. Liu, C.J. Vale, Criteria for two-dimensional kinematics in an interacting Fermi gas. Phys. Rev. A **93**, 011603 (2016)
11. J.K. Freericks, M.M. Maśka, A. Hu, T.M. Hanna, C.J. Williams, P.S. Julienne, R. Lemański, Improving the efficiency of ultracold dipolar molecule formation by first loading onto an optical lattice. Phys. Rev. A **81**, 011605 (2010)
12. S. Gopalakrishnan, I. Martin, E.A. Demler, Quantum quasicrystals of spin-orbit-coupled dipolar bosons. Phys. Rev. Lett. **111**, 185304 (2013)
13. M. Greiner, O. Mandel, T. Esslinger, T.W. Hansch, I. Bloch. Quantum phase transition from a superfluid to a Mott insulator in a gas of ultracold atoms. Nature **415**(6867), 39–44 (2002)
14. R. Grimm, M. Weidemller, Y.B. Ovchinnikov, Optical dipole traps for neutral atoms. Adv. Atom. Mol. Opt. Phys. **42**, 95–170 (2000)
15. E. Haller, M.J. Mark, R. Hart, J.G. Danzl, L. Reichsöllner, V. Melezhik, P. Schmelcher, H.-C. Nägerl, Confinement-induced resonances in low-dimensional quantum systems. Phys. Rev. Lett. **104**, 153203 (2010)
16. K.R.A. Hazzard, B. Gadway, M. Foss-Feig, B. Yan, S.A. Moses, J.P. Covey, N.Y. Yao, M.D. Lukin, J. Ye, D.S. Jin, A.M. Rey, Many-body dynamics of dipolar molecules in an optical lattice. Phys. Rev. Lett. **113**, 195302 (2014)
17. J. Heinze, S. Götze, J.S. Krauser, B. Hundt, N. Fläschner, D.-S. Lühmann, C. Becker, K. Sengstock, Multiband spectroscopy of ultracold fermions: observation of reduced tunneling in attractive Bose-Fermi mixtures. Phys. Rev. Lett. **107**, 135303 (2011)
18. N.R. Hutzler, L.R. Liu, Y. Yu, K.-K. Ni, Eliminating light shifts in single-atom optical traps (2016). Arxiv:1605.09422v1
19. R. Jördens, N. Strohmaier, K. Günter, H. Moritz, T. Esslinger, A mott insulator of fermionic atoms in an optical lattice. Nature **455**, 204–207 (2008)
20. L.R. Liu, J.T. Zhang, Y. Yu, N.R. Hutzler, T. Rosenband, K.-K. Ni, Ultracold molecular assembly (2017). Arxiv:1605.09422v1

21. M. Lu, N.Q. Burdick, B.L. Lev, Quantum degenerate dipolar Fermi gas. Phys. Rev. Lett. **108**, 215301 (2012)
22. S.A. Moses, J.P. Covey, M.T. Miecnikowski, B. Yan, B. Gadway, J. Ye, D.S. Jin, Creation of a low-entropy quantum gas of polar molecules in an optical lattice. Science **350**(6261), 659–662 (2015)
23. S. Ospelkaus, K.-K. Ni, D. Wang, M.H.G. de Miranda, B. Neyenhuis, G. Quéméner, P.S. Julienne, J.L. Bohn, D.S. Jin, J. Ye, Quantum-state controlled chemical reactions of ultracold potassium-rubidium molecules. Science **327**(5967), 853–857 (2010)
24. S. Ospelkaus-Schwarzer, Quantum degenerate Fermi-Bose mixtures of ^{40}k and ^{87}rb in 3d optical lattices. PhD thesis, Universität Hamburg, 2006
25. G. Quéméner, P.S. Julienne, Ultracold molecules under control! Chem. Rev. **112**(9), 4949–5011 (2012)
26. C.A. Regal, Experimental realization of BCS-BEC crossover physics with a Fermi gas of atoms. PhD thesis, University of Colorado, Boulder, 2006
27. L. Reichsöllner, A. Schindewolf, T. Takekoshi, R. Grimm, H.-C. Nägerl, Quantum engineering of a low-entropy gas of heteronuclear bosonic molecules in an optical lattice (2016). Arxiv:1607.06536v1
28. A. Safavi-Naini, M.L. Wall, A.M. Rey, Role of interspecies interactions in the preparation of a low-entropy gas of polar molecules in a lattice. Phys. Rev. A **92**, 063416 (2015)
29. U. Schneider, L. Hackermller, S. Will, Th. Best, I. Bloch, T. A. Costi, R. W. Helmes, D. Rasch, A. Rosch, Metallic and insulating phases of repulsively interacting fermions in a 3d optical lattice. Science **322**(5907), 1520–1525 (2008)
30. M. Schreiber, S.S. Hodgman, P. Bordia, H.P. Lüschen, M.H. Fischer, R. Vosk, E. Altman, U. Schneider, I. Bloch, Observation of many-body localization of interacting fermions in a quasirandom optical lattice. Science **349**(6250), 842–845 (2015)
31. J. Smith, A. Lee, P. Richerme, B. Neyenhuis, P.W. Hess, P. Hauke, M. Heyl, D.A. Huse, C. Monroe, Many-body localization in a quantum simulator with programmable random disorder. Nat. Phys. **12**, 907–911 (2016)
32. D. Stauffer, A. Aharon, *Introduction to Percolation Theory*, 2nd edn. (Taylor and Francis, London, 1994)
33. N. Strohmaier, D. Greif, R. Jördens, L. Tarruell, H. Moritz, T. Esslinger, R. Sensarma, D. Pekker, E. Altman, E. Demler, Observation of elastic doublon decay in the Fermi-Hubbard model. Phys. Rev. Lett. **104**, 080401 (2010)
34. T. Volz, N. Syassen, D.M. Bauer, E. Hansis, S. Dürr, G. Rempe, Preparation of a quantum state with one molecule at each site of an optical lattice. Nat. Phys. **2**, 692–695 (2006)
35. C. Weitenberg, Single-atom resolved imaging and manipulation in an atomic mott insulator. PhD thesis, Ludwig-Maximilians-Universität München, 2011
36. R.J. Wild, Contact measurements on a strongly interacting Bose gas. PhD thesis, University of Colorado, Boulder, 2012
37. K. Winkler, G. Thalhammer, F. Lang, R. Grimm, J. Hecker Denschlag, A.J. Daley, A. Kantian, H.P. Büchler, P. Zoller, Repulsively bound atom pairs in an optical lattice. Nature **441**, 853–846 (2006)
38. B. Yan, S.A. Moses, B. Gadway, J.P. Covey, K.R.A. Hazzard, A.M. Rey, D.S. Jin, J. Ye, Observation of dipolar spin-exchange interactions with lattice-confined polar molecules. Nature **501**(7468), 521–525 (2013)
39. B. Zhu, G. Quéméner, A.M. Rey, M.J. Holland, Evaporative cooling of reactive polar molecules confined in a two-dimensional geometry. Phys. Rev. A **88**, 063405 (2013)

Chapter 7
The New Apparatus: Enhanced Optical and Electric Manipulation of Ultracold Polar Molecules

In this chapter, I describe our approach to combining all the tools together that are needed for large electric fields with high-resolution detection and addressing. I begin by describing the electrode system, which allows for large, stable, homogeneous electric fields that simultaneously allow arbitrary gradients in two dimensions (2D). I go on to describe how we can couple AC microwave frequencies onto the electrodes to drive rotational transitions with high polarization fidelity. I then describe the high-resolution imaging system.

Ultracold polar molecules have long been heralded as an excellent toolbox for probing many-body long-range interacting systems [4, 20] as well as quantum information [9]. Recently, dipolar spin–exchange interactions and many-body dynamics have been observed with fermionic KRb molecules in an optical lattice [15, 36] as described in Chap. 5, and low-entropy samples in a lattice have been realized for KRb ground-state molecules [6, 21] (described in Chap. 6) and RbCs weakly bound Feshbach molecules [26]. Future work towards realizing the XXZ model of quantum magnetism will require a large DC electric field to control the strength of the Ising interaction [12, 16, 34]. Moreover, studying many-body non-equilibrium dynamics will require high-resolution in situ detection and addressing [35].

Several groups are working to implement these tools in experiments with polar molecules [11, 13], although there are few published results to date. Further, there are no published results to our knowledge that have yet demonstrated ultracold polar molecules in such an advanced apparatus. The challenges of implementing such an apparatus include the already complex nature of ultracold polar molecules which often require large magnetic fields to reach a Feshbach resonance [5, 38], combined with the seemingly conflicting requirements of large, versatile electric fields and high-resolution optical detection and addressing.

In terms of the fields required to nearly saturate the dipole moments of bialkalis, KRb is particularly demanding. For KRb, $d = 0.566$ Debye [23], but another

© Springer Nature Switzerland AG 2018
J. P. Covey, *Enhanced Optical and Electric Manipulation of a Quantum Gas of KRb Molecules*, Springer Theses, https://doi.org/10.1007/978-3-319-98107-9_7

Fig. 7.1 The dipole moments of the bialkalis as a function of the electric field. The top plot shows reactive species, and the bottom plot shows nonreactive species. Reproduced from [25]

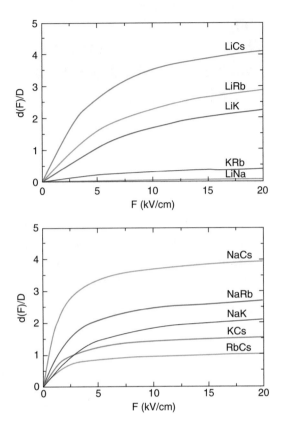

figure of merit is B/d, where B is the rotational constant, which describes the characteristic electric field. For KRb this value is $\approx 4\,\mathrm{kV/cm}$, while for, e.g., NaK it is $\approx 2\,\mathrm{kV/cm}$ [24, 33], which is thus easier to polarize. Concomitantly, to reach $>80\%$ of d for KRb, a field of ≈ 20–$30\,\mathrm{kV/cm}$ is required. Most of the other molecules being pursued have larger d and smaller B/d, so that they can reach larger dipole moments at smaller fields [25]. This is true for NaRb, NaK, RbCs, KCs, and NaCs, as shown in Fig. 7.1.

7.1 Electrode Design for Large, Versatile Electric Fields

While there has been an enormous amount of work in the cold molecule community with large electric fields [1, 2, 32], designs that are compatible with ultracold atom technology and high-resolution optical control remain unexplored. In order to reach

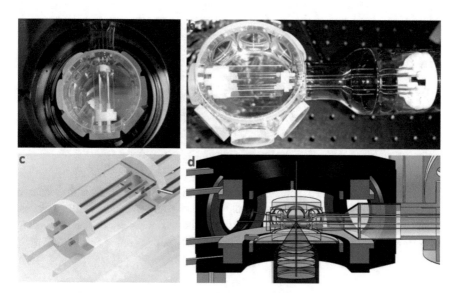

Fig. 7.2 The new apparatus KRb machine: (**a**) shows the cell with electrodes inside and is surrounded by the coils needed for atom manipulation and Feshbach association. (**b**) shows the cell alone, and the electrodes and macor holders can be seen extending further into the chamber. (**c**) shows a close-up render of a solid model of the electrodes and the macor insulators. (**d**) shows a section view of a solid model of the entire chamber and coils. The high-resolution objective below the cell is also apparent

such large electric fields in a manner than allows both homogeneous fields as well as arbitrary gradients while still providing excellent optical access, our design is based on four tungsten rods between two Indium Tin Oxide (ITO)-coated fused silica plates. This electrode assembly is located inside a fused silica octagonal cell (radius $r = 20$ mm and height $h = 35$ mm) whose windows are anti-reflection coated on both sides. The cell is surrounded by hollow-core water-cooled coils for the large magnetic fields (550 G) required to access the K-Rb Feshbach resonance [38] (see Fig. 7.2a). Figure 7.2b, c shows the electrodes in greater detail, with and without the cell. Below the cell is a high-resolution microscope objective, as shown in Fig. 7.2d.

Transparent plate electrodes are required to allow high-resolution imaging as well as the propagation of optical trapping and atom/molecule addressing beams in the vertical direction. The thickness of the ITO coating is ≈ 10 nm, which has an optical absorption at a wavelength of $\lambda = 1064$ nm of $\approx 2\%$ and a coating resistance of ≈ 1 k$\Omega/\square$, where $\square$ means "square" and denotes the entire surface area of the coating. Tungsten was chosen for the rods because it is very hard and has a high work function, which are important for reaching large electric fields, and because it is minimally magnetic. Moreover, its coefficient of thermal expansion is very small,

Fig. 7.3 The front macor
holder in the science cell. The
hole diameter for the rods is
1 mm, and the entire diameter
is ≈15 mm. The groove on
the front side holds the
ITO-coated plates

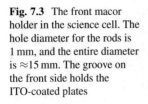

which reduces thermal expansion problems associated with baking the vacuum chamber. The electrode assembly is held together using macor insulating pieces, which were carefully designed to maximize the surface path between adjacent electrodes, and electrodes with nominally opposite sign.

7.1.1 Macor Insulators

A major challenge with this design is the macor insulators that hold the electrodes together. It is very difficult to machine such small intricate pieces and assemble the electrodes without cracking or chipping the macor. Figure 7.3 shows a solid model of the front macor electrode holder which gently rests against the front side of the science cell. The JILA instrument shop did an incredible job developing the expertise to create such pieces with a high success rate. This work was done by the late machinist Tracy Keep, and former machinist Kels Detra. None of the current JILA machinists have attempted these parts, and so certainly some expertise has been lost as a result of the untimely passing of Tracy after battling with cancer.

We anticipate reaching fields of 20–30 kV/cm, and all the macor insulators were designed to maximize the surface path length between adjacent electrodes. While the bulk dielectric strength of macor (500–1000 kV/cm) is much larger than any field we could apply, pathways on the surface that allow current to flow develop at much lower fields. Indeed, these phenomena will actually limit the fields we can reach, and a general rule of thumb is that the surface path length should be ≈10× the distance between the electrodes [28]. Note that our electrode design requires us to place the electrodes in the region of the largest field. Other high-field experiments like Stark decellerators typically try to avoid this and often have metal rods that come together from distant supports [28]. Nevertheless, even Stark decellerators are often limited by this type of breakdown.

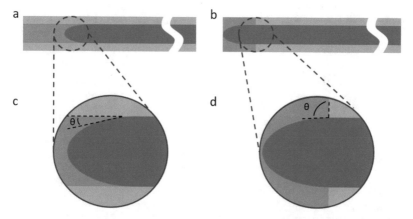

Fig. 7.4 Triple points at the end of the rod electrodes. Blue is the metal conductor (electrode), green is the dielectric (macor insulator), and orange is vacuum. (**a**) The macor sleeve extending beyond the end of the rod electrode. (**b**) The macor sleeve ending before the conical taper of the rod electrode. (**c**) A zoom in of the triple point when the macor exceeds the length of the rod. The triple-point angle is very small. (**d**) A zoom in of the triple angle when the macor sleeve is shorter than the rod. The triple-point angle is ≈90°

7.1.2 Triple Points

Another important consideration while designing electrodes for large electric fields is triple points, which are defined as the junction of conductor, dielectric, and vacuum. Since the electric field inside a perfect conductor is identically zero, there is a discontinuity at the surface of the electrodes (see Fig. 7.7c), and the addition of a dielectric material can add substantial complications. These triple points serve as electron sources and are generally regarded as the location where flashover is initiated in the insulators [18]. They are also a vulnerable location from which radio-frequency (rf) breakdown is triggered [18].

This issue will be discussed more in Chap. 8, but the most acute area for triple-point-induced breakdown on our electrodes is at the end of the rods inside the macor sleeves. These problems are illustrated in Fig. 7.4, and two cases are considered. If the macor sleeve is longer than the rod as shown in Fig. 7.4a, then the triple-point angle θ is very small as shown in Fig. 7.4c. However, if the rod extends beyond the macor sleeve as in Fig. 7.4b, then the triple-point angle is $\theta \approx 90°$. The latter case is much better for minimizing the electron density in the dielectric material and thus reduces the propensity to generate an electron beam from the electrode or facilitate rf breakdown. This hypothesis has been verified with measurements in the new apparatus, and will be discussed in Chap. 8.

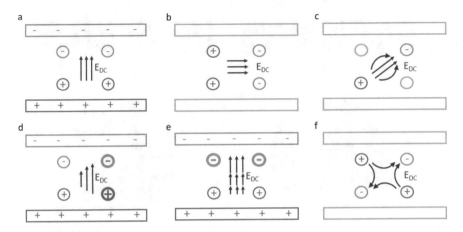

Fig. 7.5 The electric field orientations that are possible with the electrode configuration of the new apparatus. (**a**) shows how large, homogeneous fields in the vertical direction can be generated. (**b**) and (**c**) show how this field can be tilted to an arbitrary angle in the 2D plane. Note that the field homogeneity is reduced in these orientations. (**d**) and (**e**) show how arbitrary gradients can be applied in any direction in the 2D plane. (**f**) shows a quadrupole field which can be used to trap molecules in 2D

7.2   The Flexibility of the Electrode System

Such an electrode configuration offers many useful electric field orientations, as depicted in Fig. 7.5. Figure 7.5a shows how a homogeneous electric field in the vertical direction can be generated, and will next be discussed in detail. The direction of this field is highly adjustable, as shown in Fig. 7.5b, c (although the field is less uniform when oriented at an angle), which is useful for tuning the dipole angle relative to the collision angle (see, e.g., conceptually similar experiments with highly magnetic atoms [10, 19]). Further, Fig. 7.5d, e shows how arbitrary gradients can be applied in the plane perpendicular to the long direction of the electrodes. Lastly, a quadrupole configuration can be used, as shown in Fig. 7.5f. Any combination of these six orientations is also possible, yielding enormous flexibility.

7.3   Homogeneous Electric Fields

As stated above, homogeneous electric fields are important for realizing the XXZ model of quantum magnetism. This is manifest upon consideration of the strength of the spin–exchange term $J_\perp/2$ (see, e.g., [34]), which for KRb is 100 Hz between neighboring sites in a lattice of spacing $\lambda/2 = 532$ nm [15, 36]. In order for spin–exchange processes to be resonant between distant sites, the energy offset between

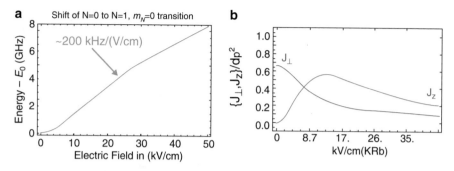

Fig. 7.6 The energy shift of the $|0, 0\rangle$ to $|1, 0\rangle$ transition as a function of the electric field (**a**), and the strength of $J_\perp$ and J_z as a function of field (**b**). Adapted from [16]

them must be $\sim$10's of Hz. Any electric field gradient or curvature will limit the length scale of these interactions. A figure of merit specifically from recent KRb experiments in an optical lattice is the energy spread across the cloud due to residual differential AC Stark shifts from the optical traps (discussed in Chap. 5) which were limited to $\sim$1 kHz [22, 36].

Several groups are pursuing an electrode system of just four rods, similarly aligned in a quadrupole fashion. However, this configuration is incapable of providing purely homogeneous fields. On the other hand, perfectly parallel plates will provide a homogeneous field, but are not versatile and cannot provide arbitrary gradients. The configuration presented in this work can allow homogeneous fields that are simultaneously versatile, but the relative voltages on the rods and plates must be chosen very carefully such that the curvature from the rods does not dominate the flat field contribution from the plates. The separation between the plates (whose thickness is 1 mm) is 6 mm, and the vertical separation between the rods (whose diameter is 1 mm) is 3 mm. The horizontal separation between the rods is 5 mm. Finite element analysis (FEA) using COMSOL Multiphysics was performed to study the electric field distribution for arbitrary voltages on all six electrodes.

It is constructive to understand what electric field homogeneity is needed for all the experiments that we would like to do. The zeroth-order requirement is that the electric field gradient or curvature is small compared to the optical dipole force of the trap, such that applying a field does not spill molecules out of the optical trap. A higher-order requirement that the new apparatus was designed to meet is an energy shift between lattice sites of less than $J_\perp$. The energy shift of the $|0, 0\rangle$ to $|1, 0\rangle$ transition as a function of the electric field is shown in Fig. 7.6a. For a cloud of $\approx$100 μm ($n_{tot} \approx 200$ sites), the total shift across the cloud $n_{tot} J_\perp \approx 40$ kHz, for which we require a gradient of $E' < 20$ V/cm^2. At a field of $E = 20$ kV/cm, this level of homogeneity corresponds to $1/10^5$ across the cloud.

Figure 7.7a, c shows cuts of the magnitude of the electric field when the voltages on the plates are ± 10 kV and the rods are varied around ± 4.225 kV, for which the electric field is 32 kV/cm. Figure 7.7a shows the horizontal cut, and the cat ears

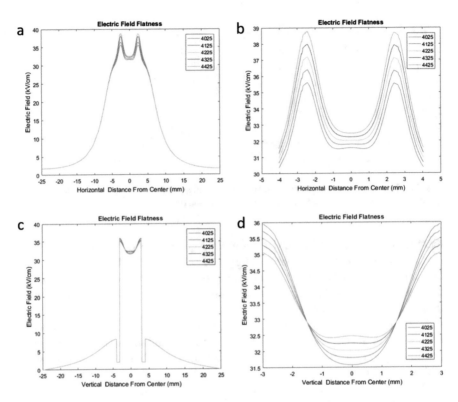

Fig. 7.7 Simulated electric field cuts along the axial and radial directions. (**a**) A cut through the origin along the axial direction (through the plates) for the plates at $\pm 10\,\mathrm{kV}$ and the rods at $\pm 4.225\,\mathrm{kV}$. (**b**) A cut through the origin along the radial direction (between the plates) for the plates at $\pm 10\,\mathrm{kV}$ and the rods at $\pm 4.225\,\mathrm{kV}$. (**c**) A zoom in of the axial cut for several values of the rod voltage with the plates at $\pm 10\,\mathrm{kV}$. (**d**) A zoom in of the radial cut for several values of the rod voltage with the plates at $\pm 10\,\mathrm{kV}$. Reproduced from [7]

immediately between the rods are apparent. Thus, the field is not homogeneous over the entire range, but it is flat over a roughly ± 1 mm region around the center. Figure 7.7c shows the vertical cut, and a similar behavior is evident around the geometric center.

It is constructive to look at the flatness of the field near the center as a function of the voltage ratio between the rods and the plates. Figure 7.7b, d shows a zoom of the axial and radial cuts, respectively, where the rod voltage is varied around $\pm 4.225\,\mathrm{kV}$ while holding the plates at $\pm 10\,\mathrm{kV}$. The different colors show the different values of the rod voltages, and it is clear that field in the center has curvature either up or down depending on whether the rod voltage is too high or low, respectively. The purple curves show the optimal field where the field is flat over ≈ 1 mm on either side of the center. Note that the curvature is opposite in the radial and axial directions as is required by Gauss's law $\vec{\nabla} \cdot \vec{E} = 0$.

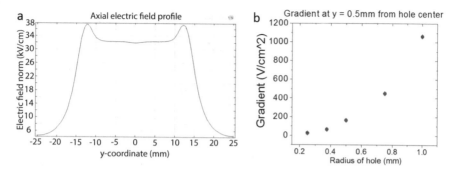

Fig. 7.8 Simulating the field effects of a hole in the ITO plate. (**a**) The field profile through the origin and along the symmetry axis of the electrodes. The shallow dip in the middle is caused by a hole in the plates. (**b**) The gradient of the field 500 μm from center of the hole as a function of the size of the hole

7.3.1 Effects from a Hole in the ITO-Coated Plates

We can also use these simulations to understand how sensitive this maximally homogeneous regime is to imperfections. The first case we consider is a hole in the ITO plate. Such a configuration is interesting since we could have alternatively used metal plates, which would require a hole for the vertical lattice, Raman beams, and a probe beam. Moreover, as discussed in Chap. 8, we initially had a lot of trouble pairing the ITO coating with an AR coating. Lastly, the ITO coating could be damaged in a very localized region. This could happen if a lot of atoms or other dirt land on a small region, or if the vertical lattice beam goes through a dirty region and locally burns something off.

Figure 7.8 shows the case of a hole centered on the electrode assembly and centered on the molecules. Figure 7.8a shows a cut through the origin of the electrodes and along the symmetry axis. The dip in the middle is due to the hole in plates. Note that the ears at the ends of the flat field region are an artifact of a difference between the lengths of the plates versus the rods. To quantify the effect of this hole, we consider the gradient of the field slightly off axis of the center of the hole, by 500 μm. We can look at this gradient as a function of the size of the hole, as shown in Fig. 7.8b.

The scaling is rather unfavorable, and even a hole of radius $R = 500$ μm causes a gradient of ≈ 150 V/cm^2 at a field of ≈ 30 kV/cm. Note that the energy gradient due to the variation in the DC Stark shift is comparable to the optical dipole force in the radial direction of typical optical dipole traps, and so such gradients will severely perturb or even spill the molecules out of an optical trap. A hole of this size is required to safely propagate typical laser beams, and so this scheme is not practical. Note, however, that if the hole is very small or far away from the molecules, these effects are mitigated enormously. Therefore, we expect the effects of localized damage or patch charges to be negligible.

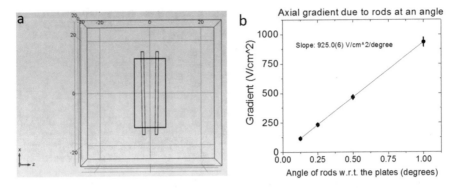

Fig. 7.9 Electric field gradients induced by a wedge angle between the rods. (**a**) The geometry of the simulation, where many wedge angles were simulated. (**b**) The gradient of the field at the origin as a function of the wedge angle

7.3.2   The Effect of an Angle Between the Rods and/or Plates

To understand whether it is possible to reach 10^{-5}-level field homogeneity, it is also important to consider the parallelism of the plates and rods, which can be roughly estimated without the use of FEA. If the spacing between two rods or plates are d and d' on the two ends separated by r, then the gradient is roughly $E' = (\Delta V/d - \Delta V/d')/r = \Delta V/d \cdot \Delta d/(d \cdot r) = E \Delta d/(d \cdot r)$, where $\Delta d = d - d'$. Note that $\Delta d/r$ is also the wedge angle θ. For $E = 30\,\text{kV/cm}$, $d = 6\,\text{mm}$, and $r = 20\,\text{mm}$, limiting the gradient to $E' = 20\,\text{V/cm}^2$ requires $\Delta d \approx 10\,\mu\text{m}$. To confirm this, Fig. 7.9 shows the effect of a wedge angle between the rods and plots the gradient as a function of the angle.

Figure 7.9b shows that reaching a gradient of $20\,\text{V/cm}^2$ requires a wedge angle less than $\theta = 0.02°$, which corresponds to $\Delta d = 7\,\mu\text{m}$. While this is not impossible, it will certainly be difficult to realize in practice. However, the situation becomes much more favorable at lower fields. Firstly, the gradient scales as E, so we win linearly there. More importantly, however, the slope of the energy shift decreases as the dipole moment decreases. Therefore, at a field of $E \approx 10\,\text{kV/cm}$ we expect this level of homogeneity to be possible. Figure 7.6b shows the strength of $J_{\perp}$ and J_z as a function of the electric field. This shows that J_z is actually maximized at around this value [16], so there will likely not be any reason to try to do quantum magnetism in larger fields.

7.3.3   Charging Scenarios

Chapters 2 and 6 described the severe problems in the old apparatus with the build-up of transient charges on the glass cell. Hence, it is very important that such issues do not plague our work with electric fields in the new apparatus. To estimate the significance of these effects, we consider putting potentials of $\pm 1\,kV$ on various dielectric materials in the chamber, such as the glass cell or the macor insulators. We then look at the field profile when all the electrodes are grounded. We simulated the effects of potentials in three places: (1) the edges of ITO-coated plates, or the edges of the windows where they are sealed to the quartz frame of the cell, specifically the edges of the side and top and bottom windows which are nearest to the ITO-coated plates; (2) the end window and its edges; and (3) the macor holders, which are the surfaces nearest to the molecules.

First note that the electrode assembly acts as an effective Faraday cage, which blocks electric fields from distant objects. In case (1), the potential applied to the edges of the ITO-coated plates or the edges of the windows gives rise to a field at the center of the electrodes of ≈ 1 V/cm and is flat over roughly the separation of the rods, which is 4 mm. Note that the shielding from the electrodes is the best along this radial axis. In the case of (2) where potential is put on the end window, the symmetry of the electrodes along this axis does not allow us to shim any stray fields. Fortunately, the end window is far enough away that the electric field from this potential has fallen to also ≈ 1 V/cm at the center of the science cell. The gradient of this field is also very small, ≈ 1 V/cm^2. Case (3), where potentials are placed on the macor insulators, is by far the most concerning scenario. These potentials give rise to fields at the center of ≈ 20 V/cm, but this of course depends on the potential. $\pm 1\,kV$ is perhaps unrealistically high, but hopefully this will give us an idea of what to expect in the worst case. While this field is larger than the other cases, it is still fairly uniform at the center. Nevertheless, we will ultimately be limited by such effects, as well as patch charge build-up on the ITO-coated plates, as discussed above.

7.4 Arbitrary Electric Field Gradients in 2D

While a homogeneous field is ideal for studying spin models with polar molecules, gradients are essential for pursuing other directions. For instance, consider an array of 2D gases in a 1D optical lattice which is propagating in the vertical direction (through the ITO-coated plates). Applying the field along the vertical direction stabilizes the molecules through the quantum stereodynamics of their collisions [8]. Then, a gradient in the horizontal direction can be used to tilt the 2D traps for direct evaporative cooling of the molecular gas [37], which is a promising route to create the first bulk quantum degenerate gas of polar molecules as discussed in Chap. 6. Further, a vertical gradient can be used to create a unique DC Stark shift of a rotational transition for each 2D subsystem of the 1D lattice. This can be used to spectroscopically select a single layer, as is done routinely using the Zeeman shift for atoms [27], which is important for high-resolution in situ detection.

We can start to add a gradient of the electric field in any arbitrary direction in 2D. This is done by biasing some of the rods away from the voltage ratio between the

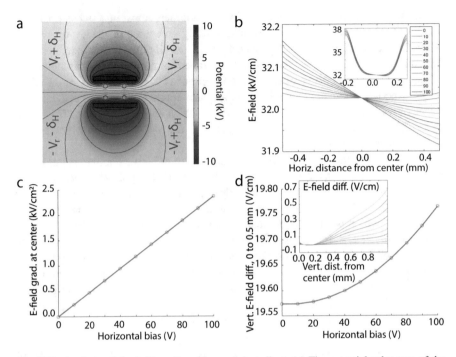

Fig. 7.10 Analysis of the field profile with a radial gradient. (**a**) The potential color map of the electrodes when the plates are at $\pm 10\,\mathrm{kV}$ and the rods are at $V_r = 4.225\,\mathrm{kV}$ with some bias δ_H along the horizontal (radial) direction. The lines are equipotentials, and the color map is within $\pm 10\,\mathrm{kV}$. (**b**) The electric field at the center as a function of the horizontal position. The colors represent different values of δ_H between 0 and 100 V. The inset shows a zoomed out view where the peaks on the edges are the locations of the rods, and the voltage on the left rod is larger than the voltage on the right by $2\delta_H$. (**c**) The gradient of the electric field in the radial direction as a function of δ_H. (**d**) The field variation between the center and 0.5 mm in the vertical direction as a function of δ_H. The inset shows the field variation in the vertical direction as a function of the position between 0 and 1 mm. The different colors denote the different values of δ_H between 0 and 100 V. Reproduced from [7]

rods and plates of 0.4225 that is nominally used for the homogeneous field described above. Figures 7.10 and 7.11 show the gradient of the field as a function of the voltage bias δ in the horizontal (radial) and vertical (axial) directions, respectively. Typical gradients required in the horizontal direction for evaporation give energy shifts similar to the radial trap frequency, which are $\approx 100\,\mathrm{V/cm^2}$ for KRb at large electric fields [37]. Figure 7.10c shows that such a gradient can be generated with $\delta_H = 8\,\mathrm{V}$, which is easily controllable with typical voltage stability.

For selection of a single layer in the 1D lattice, energy shifts of $\approx 10\,\mathrm{kHz/site}$ have been used for ultracold atoms [27]. For KRb at large electric fields ($> 30\,\mathrm{kV/cm}$), the typical energy shifts are $\approx 100\,\mathrm{kHz/(V/cm)}$ (see Fig. 7.6a). Figure 7.11b, d shows the sensitivity to a bias of the rods in the vertical direction, which suggests that a bias of $\delta_V \approx 200\,\mathrm{V}$ is sufficient to give a 10-kHz energy shift (2 kV/cm^2 needed

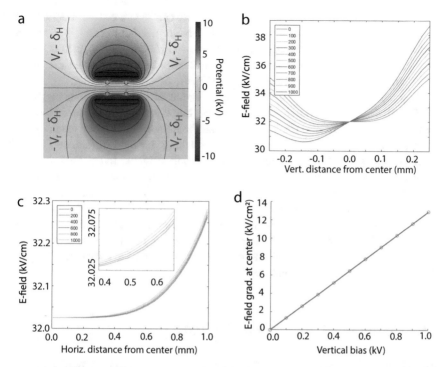

Fig. 7.11 Analysis of the field profile with an axial gradient. (**a**) The potential color map of the electrodes when the plates are at ± 10 kV and the rods are at $V_r = 4.225$ kV with some bias δ_V along the vertical (axial) direction. The lines are equipotentials, and the color map is within ± 10 kV. (**b**) The electric field around ± 2.5 mm of the center in the vertical direction. The different colors denote different values of δ_V between 0 and 1000 V. (**c**) The electric field as a function of the horizontal position between the center and 1 mm. The different colors denote the same range of δ_V between 0 and 1000 V. The inset shows a zoom around 0.5 mm. (**d**) The gradient of the electric field in the axial direction as a function of δ_V. Reproduced from [7]

to separate adjacent layers of spacing $\lambda/2 = 532$ nm by 0.1 V/cm). Moreover, in order to select individual layers of the 1D lattice with high fidelity, the field strength will have to be stable enough such that field variation at the layer during the pulse is much smaller than the field difference between layers (0.1 V/cm). We have thus designed high voltage servos for electrodes that are expected to operate at the several parts per million level, giving ≈ 10 mV stability and precision. This corresponds to fluctuations of the order 0.1 V/cm.

In fact, layer selection becomes easier at lower electric fields where the absolute field fluctuation is pushed down. However, by operating at a lower electric field, the dipole moment is still in the linear regime, and can easily be tuned between 0–200 kHz/(V/cm). Thus, the field gradient will need to be even larger to reach the same energy shift between layers. Fortunately, the electrode geometry allows for very large vertical gradients, and the plot shown in Fig. 7.11d continues up to ~ 15 kV/cm^2. For concreteness, consider working at 5 kV/cm where the energy shift is

≈ 50 kHz/(V/cm). An energy shift of 10 kHz between layers can be reached with ≈ 10 kV/cm^2 gradients, induced by $\delta_V \approx 1$ kV. Note that the bias δ_V is becoming similar to the voltages nominally applied to the rods for the homogeneous field at 5 kV/cm (i.e., $\approx 4.225/6$ kV). Here, the field difference between adjacent sites is still 0.1 V/cm, but the absolute field stability is 10 mV/cm.

Figure 7.11c shows the curvature in the radial direction due to Gauss's law $\vec{\nabla} \cdot \vec{E} = 0$. We require the field variation across one layer to be well within the Rabi frequency $\Omega \approx 10$ kHz, and much smaller than the field difference between adjacent layers. Even for the largest bias $\delta_V = 1$ kV where the field difference between adjacent layers is 1 V/cm, the electric field variation in the radial direction near the center is smaller than the numerical error, $\ll 1$ V/cm. Note that it is important to work in the maximally flat field configuration $V_r/V_p = 0.4225$, and that it is important that the cloud be within $\approx 200\,\mu$m from the center in the horizontal direction especially. Additionally, it is important that the bias δ is divided evenly between each pair of rods $\pm \delta/2$, as in Figs. 7.10a and 7.11a. This is essential for maintaining the homogeneity of the field and minimizing the curvature along the direction opposite to the applied gradient.

7.5   Coupling AC Fields onto the Rods

We also designed our apparatus for coupling AC microwave frequencies onto the rods, which we can use to directly drive rotational transitions. The rods are separated by several mm's, which is significantly smaller than the wavelength of the microwave field in the frequency range of 2–8 GHz. Therefore, the microwave field can be thought of as a homogeneous AC electric field on the rods. We have the ability to change the relative angle between the AC and DC electric fields, which gives us the ability to control the polarization of the microwave field and thus the angular momentum of the rotational transitions.

Figure 7.12 shows the elements of the electrode system that are required to couple microwaves onto the rods. A bias tee of high voltage inductors and capacitors are used to independently couple in DC and AC voltages, respectively [29]. The schematic is shown in Fig. 7.12a, and the arrangement of these elements is shown in b&c. The inductors are simply two loops of Stainless Steel 316 (SS316) wire, and the inductance was chosen to maximize the transmission of microwave signals of frequency 2–8 GHz. The capacitors are shown in c, and the top capacitor plates are comprised of SS316 pillars. The disks at the bottom of the pillars sit on a macor base, whose thickness of 1 mm serves as the dielectric material of the capacitor. Below the macor is another such disk which continues below to a four-rod low-power feedthrough.

This feedthrough is not impedance matched with a coaxial input, but it allows us to change from outside the chamber how the microwaves are coupled to the four rods, giving us control of polarization. It is imperative that this bias tee is placed inside the vacuum chamber because the DC electric field across the capacitor plates

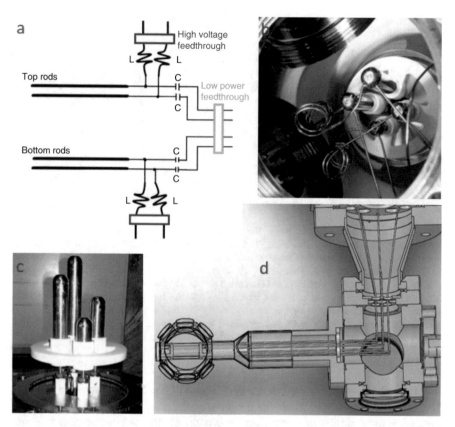

Fig. 7.12 The electrodes inside the new apparatus. (**a**) shows a schematic of the bias tee and how AC fields are coupled into the chamber. (**b**) shows the bias tee in the large cube inside the vacuum chamber. (**c**) shows the home-made capacitors used to couple the AC microwave fields onto the rod electrodes in the science cell. (**d**) shows the transmission line from the large cube into the science cell

will be up to 100 kV/mm when 10 kV are applied to the rods, and this field is much larger than the dielectric breakdown of air. We measure that the insertion loss going from a coaxial cable below the chamber through the capacitors is ≈ 5 dB. This will be discussed more in Chap. 8.

The capacitance of the coupling capacitors were chosen based on an estimate of the capacitance between the rods in the science cell, which is <1 pF. The capacitance of the coupling capacitors can thus be thought of as an AC voltage divider, and since the impedance of a capacitor is $Z_C = 1/(i\omega C)$, we want the capacitance of the coupling capacitor to be larger than that of the rods. Thus, the capacitance was designed to be ≈ 4 pF. Figure 7.12d shows the transmission line from the feedthroughs in the large cube into the science cell. This is also shown in Fig. 7.13. We measure ≈ 20 dB transmission loss of microwaves from the coupling capacitor

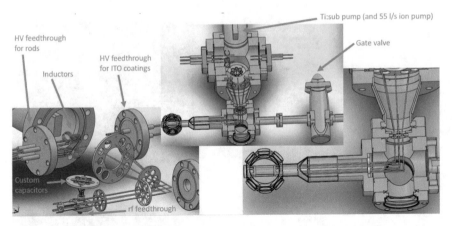

Fig. 7.13 The electrode layout inside the science chamber. The large cube that houses the feedthrough for the AC and DC voltages is connected to the pumping region of the chamber, but also the electrode path to the science cell. The cube contains four inductors and four capacitors to decouple the AC and DC signals. Six rods go from the large cube all the way to the science cell, where two of the rods connect to ITO-coated plates. The four-channel feedthrough connects high voltage through the inductors that ultimately go to the rods in the science cell. The two-channel feedthrough connects high voltage ultimately to the plates in the science cell. There are no inductors because the plates are not connected to capacitors or high frequency feedthroughs. We opted to use a four-pin low-power feedthrough for coupling in the microwaves so that we can control outside the chamber to which rods are connected

into the rod electrodes in the science cell. This will be discussed more in Chap. 8. Despite these losses, we expect the coupling of the microwave electric field to the molecules to be similar or larger than the broadband microwave horn that we used in the past because the rods are so much closer and there are no longer any grounded magnetic field coils between the molecules and source.

7.6   High-Resolution Imaging System

The rods are arranged within the plates in a rectangular way in order to accommodate an optical numerical aperture (NA) in the vertical direction as high as 0.75. In practice, the imaging resolution will be limited by the objective as well as the optical quality of the intermediate surfaces. The RMS surface flatness before assembly and baking over the entire aperture of the large, vertically oriented windows of the cell is $\lambda/5$ at $\lambda = 780$ nm. The ITO-coated plates are expected to have better flatness, but slightly worse surface figure. The electrode configuration of the rods and plates was carefully oriented to be as parallel to the large cell windows as possible. Any angle will likely limit the ultimate NA that can be achieved, and an imaging system that corrects the coma aberrations induced by this angle would likely be needed for high NA objectives. The working distance of the objective is 22 mm, and the optical path

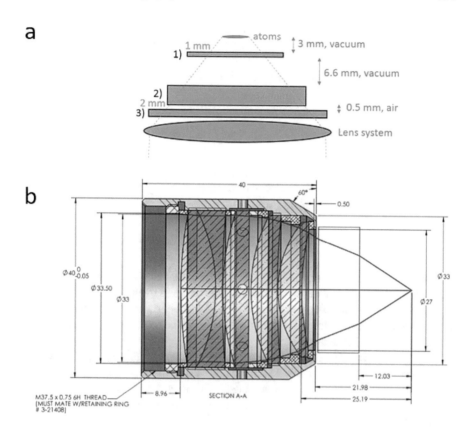

Fig. 7.14 (**a**) The optical path between the atoms/molecules to the objective. The thicknesses of the substrates are indicated in green. (**b**) The design of the objective, accounting for all the optical elements shown in (**a**). The optical elements are outlined in purple, including the fused silica substrates. Courtesy of Special Optics

between the atoms/molecules and the objective is shown in Fig. 7.14a. The design of the objective is shown in Fig. 7.14b.

The microscope objective shown in Fig. 7.2d is designed to have NA = 0.53, correcting for the spherical aberration of the ITO-coated plate and the cell window (whose thickness is 6.4 mm). The coma induced by the tilted ITO-coated plates is not corrected, but we expect its effect to be relatively small for a 0.1–0.2° angle at this NA. The diffraction-limited Rayleigh spot size is R_{min} = 880 nm, and such spot sizes have been reached in testing. The ITO-coated plate is paired with a broadband anti-reflection coating from λ = 700–1100 nm, which is designed to minimize reflection at 1064 nm in order to accommodate a vertical optical lattice beam. The cell window is coated over a similar bandwidth, and also optimized at 1064 nm. In order to reflect the vertical lattice beam, which propagates from above, before going through the objective, a UVFS dichroic filter plate of 2 mm thickness is placed below the cell. Figure 7.2d shows this filter, which can be adjusted with a 2D kinematic mount. The microscope objective is inside the mount for this filter, and can be independently adjusted with a five-axis stage.

7.6.1   Objective Design and Aberration Theory

It is constructive to consider the mathematical structure of the primary (called Seidel) aberrations, where the wave aberration is given by the Zernike polynomials [3, 14]:

$$\phi^{(4)} = -\frac{1}{4}B\rho^4 - Cy_0\rho^2\cos^2\theta - \frac{1}{2}Dy_0^2\rho^2 + Ey_0^3\rho\cos\theta + Fy_0\rho^3\cos\theta, \quad (7.1)$$

where ρ is the radial coordinate in cylindrical coordinates, θ is the polar angle, and y_0 is an offset from the center of the imaging path defined by the lenses. These terms are specific aberrations: the first term B is spherical aberration, the second term C is astigmatism, the third term D is curvature of field, the fourth term E is distortion, and the last term F is coma [3].

The substrates between the objective and the object plane cause spherical aberrations, which must be corrected in the objective. This is done using a meniscus lens as the first lens in the five-optic objective lens system, which can be seen in Figs. 7.2d, 7.14b, and 7.15. Since spherical aberration scales as ρ^4, an objective design where the first lens is a meniscus lens can be used for corrections. The final lens pair in Fig. 7.14b is a doublet. As discussed above, another important term in the aberration error is coma, and a possible source is the relative angle between the cell windows and the ITO-coated plates. The $\cos\theta$ scaling of this term suggests that a wedged compensation plate could be placed after the objective. However, since it also scales as ρ^3 this wedged plate would likely need a polynomial cylindrically symmetric taper in addition to the wedge.

The objective is designed only for $\lambda = 767$ and $\lambda = 780$ nm, and thus substantial chromatic aberrations are expected for significantly different wavelengths. This will be important for the high-resolution addressing beams and holographic digital micromirror device (DMD) projection onto the atoms discussed in Chap. 10. The wavelengths that we intend to use for this are $\lambda = 690$ nm and $\lambda = 1064$ nm. We expect that the effects of shorter wavelength (smaller R_{min}) and chromatic aberration (larger R_{min}) roughly cancel for $\lambda = 690$ nm, allowing us to reach a Gaussian $1/e^2$ radius for an addressing beam of ≈ 750 nm. $\lambda = 1064$ nm is both further away from $\lambda = 780$ nm, and has a longer wavelength. Accordingly, we expect the minimum waist of an addressing beam with this wavelength to be roughly 1 μm, which is still sufficiently small to allow addressing at the two-site level in the lattice.

Fig. 7.15 The coils and objective path around the science cell. The objective is shown below the science cell, and around is the holder for the dichroic plate that goes between the science cell and the objective. The objective is on a five-axis stage, and the dichroic is separately on a two-axis kinematic mount

Fig. 7.16 Photographs of the coil mount and objective and dichroic mounts around the science cell. The objective is on a five-axis mount that is inside and separate from the tube of the two-axis kinematic mount which holds the dichroic filter. This filter is necessary to reflect the vertical lattice beam before the objective

7.6.2 The Coil and Optics Mounts Around the Science Cell

The coils and objective path around the science cell are shown in Fig. 7.15. The objective is shown below the science cell, and around it is the holder for the dichroic plate that goes between the science cell and the objective. This dichroic filter is used to retroreflect the vertical lattice beam, while the Raman beams and probe beams propagate through the objective (to be discussed more in Chap. 9). The objective is on a five-axis stage, and the dichroic is separately on a two-axis kinematic mount. Photos of how this looks in the lab are shown in Fig. 7.16. The objective and dichroic holders are made of ultem, and most of the other components are made of aluminum (which is anodized). The black aluminum plate has a slit cut in it to prevent eddy currents when the vertically oriented fields switch quickly. The coil mount is made of G-11 glass/epoxy phenolic, and then painted black with a low-outgassing paint. The legs for the coil mount and the objective and dichroic mounts are titanium in an effort to reduce magnetization of nearby materials.

7.6.3   Deflection of the Science Cell Windows Due to the Differential Pressure

It is also constructive to consider the deflection of the windows on the vacuum chamber due to the differential pressure. Continuum mechanics provides Kirchhoff–Love plate theory, which describes the deflection of a material (i.e., Fused Silica) due to a differential load with various types of supports. Using this analysis, we can write the deflection as a function of the distance from the center of the window [31]:

$$w(r) = \frac{q}{64 \cdot D'} \left(a^2 - r^2\right)^2, \tag{7.2}$$

where q is the differential pressure 10^5 Pa, $a \approx 20$ mm is the radius of the window, and $D' = 2h^3 E/(3(1 - \nu)^2)$ is the flexural rigidity, where $2h$ is the plate thickness, E is the elastic modulus, and ν is Poisson's ratio. An interesting note from this equation is that spherical aberrations also scale as r^4, so vacuum deflections particularly exacerbate the spherical aberration which are already present from the vacuum window.

Figure 7.17 shows a finite element simulation of the deflection amount, and it suggests that the maximum deflection is $w_{max} \approx 0.2\,\mu$m, which is of order $\lambda/5$. This level is similar to the flatness due to the science cell assembly procedure and baking, and analysis of the flatness will be discussed in Chap. 8. Note that w_{max} depends strongly on a, and so larger windows will be more affected by this issue unless h is increased accordingly. However, even if h is increased the spherical aberration it causes will become more significant, and the objective design will inevitably become more complicated in order to compensate. However, increasing h will cause more spherical aberration, which complicates the design of the microscope objective.

7.7   Magnetic Fields

The coils around the science cell are shown in Figs. 7.2, 7.15, 7.16, and 7.18. The bias coils are 16 turns (4 × 4) of hollow-core square coil where water flows through every turn. Inside the bias coils and concentric with them are the

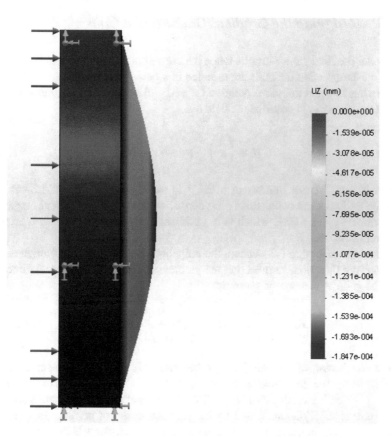

UZ (mm)

0.000e+000

-1.539e-005

-3.078e-005

-4.617e-005

-6.156e-005

-7.695e-005

-9.235e-005

-1.077e-004

-1.231e-004

-1.385e-004

-1.539e-004

-1.693e-004

-1.847e-004

Fig. 7.17 Deflection of the vacuum window due to the differential pressure. A finite element simulation of the deflection of window, where the color scale shows the deflection as described by the color bar

gradient/quantization coils, which are a 3×9 array of magnetic wire. These coils are not actively cooled, but their proximity to the bias coils provides excellent passive cooling. These coils have an H-bridge which allows us to change the current direction of one coil with respect to the other, and switch them from an anti-Helmholtz quadrupole field to a Helmholtz field which could be used for fast magnetic field sweeps, providing a quantization field in the vertical direction, or shimming out the vertical component of the stray field during optical molasses or Raman side band cooling in a quantum gas microscope (to be discussed in Chap. 11).

7.7.1   Designing the Gradient/Quantization Coils

To model the field from the coils when the current is flowing the same direction in both, we assume that the coils are oriented in a perfect Helmholtz geometry where their radius is equal to their separation ($R = d$). In our case $R > d$, but we can use this approximation to describe the field by:

$$B = \left(\frac{4}{5}\right)^{3/2} \mu_0 \cdot N \cdot I/R, \qquad (7.3)$$

where $N = 27$ is the number of turns, I is the current, and $R = 4.2$ cm is the average radius of the coils. Taking the derivative with respect to I and plugging in the numbers gives 6 G/A, which is in excellent agreement with the measurements shown in Chap. 8.

When the H-bridge is switched, the coils have current flowing through them in the opposite direction. In this case, the gradient of the field in the axial direction z near the geometric center is given by:

$$B_z = 2 \cdot \frac{3}{2}\mu_0 \cdot N \cdot I \cdot R^2 \frac{d/2}{\left((d/2)^2 + R^2\right)^{5/2}}, \qquad (7.4)$$

where the factor of two in front is because this gradient is from each coil. Again, taking the derivative with respect to I gives the gradient as a function of current, which is calculated to be ≈ 2 G/cm/A. Measurements in Chap. 8 show that this value is $2.4(1)$ G/cm/A, and so the calculation is not perfect. This could be caused by a slight error in the estimates of R and d, for which average values are used. The way this equation scales with R makes it particularly sensitive to miscalculations. Nevertheless, such a simple equation can be used for reasonable qualitative agreement.

7.7.2   Designing the Bias Coils

The bias coils were painstakingly designed such that the average separation between the 16 turns of the two coils d and the average radius of each coil R are exactly the same. Moreover, the two coils were wound in the opposite way such that the gradients from the winding imperfections of each coil would cancel each other out. The careful consideration to detail in the design of these coils is what gives rise to the incredible flatness measured in Chap. 8. The equation for the magnetic field from these coils is the same as the equation above for the quantization coils, and the field per current for the bias coils is calculated to be 2.55 G/A, which is in excellent agreement with the measurements presented in Chap. 8.

The K-Rb Feshbach resonance is at 547 G, which can be reached with ≈ 220 A. At 1005 G is a narrow Feshbach resonance for Rb, which could be useful to measure double occupancy of Rb by making Rb_2 Feshbach molecules. This field can reached with ≈ 400 A. Another important field is 1260 G, at which the excited rotational states $|1, 1\rangle$, $|1, 0\rangle$, and $|1, -1\rangle$ are degenerate [30] (discussed in Chap. 11). This field can be reached at ≈ 480 A, which is still within reach of our high-current power supply. The coils in the old apparatus had 12 turns, and roughly 330 A were required to reach the Feshbach resonance. Moreover, the old coils were not in the perfect Helmholtz configuration, and so they had a larger curvature across the cloud.

7.8 The Chamber and Vacuum Considerations

The full vacuum chamber of the new apparatus is shown in Fig. 7.18. Similar to the old apparatus, a moving quadrupole trap carries atoms from the MOT chamber (not shown) to the quadrupole evaporation region. A gate valve and a flexible bellows separate the two regions, as in the old apparatus. The evaporation region is near the small cube, attached to which is a non-evaporable getter (NEG) pump (SAES CapaciTorr D 400-2) which nominally has a pumping speed of 400 L/s. After plugged quadrupole evaporation (discussed in Chap. 9) in the region near the small cube, the K and Rb clouds are optically transferred into the science cell. Behind the large cube where the high voltage feedthroughs are shown, there is a 55 L/s ion pump and titanium-sublimation pump.

More details of the chamber will be described in Chaps. 8 and 9, and here I will present only the details necessary to analyze the gas conductance through the chamber. The pressure P in the science cell is given by $P = Q/S$, where Q is the gas load and S is the effective pumping speed. The gas load is generally complicated

Fig. 7.18 The full vacuum chamber of the new apparatus. Similar to the old apparatus, a moving quadrupole trap carries atoms from the MOT chamber (not shown) to the quadrupole evaporation region. A gate valve and a flexible bellows separate the two regions, as in the old apparatus. The evaporation region is near the small cube, attached to which is a non-evaporable getter pump. After plugged quadrupole evaporation in the region near the small cube, the K and Rb clouds are optically transferred into the science cell. Behind the large cube where the high voltage feedthroughs are shown that there is an ion pump and titanium-sublimation pump

to calculate, but it scales with the surface area SA and the outgassing rate R as $Q \sim SA \cdot R$. The effective pumping rate S is given in terms of the pumping speed from the ion pump and titanium-sublimation (Ti-sub) pump S_p, and the conductance [17]:

$$\frac{1}{S} = \frac{1}{C} + \frac{1}{S_p} \tag{7.5}$$

such that they add in parallel.

The conductance through a tube in the molecular flow limit is given by $C = 78 D^3/L$, where D is the tube diameter and L is the tube length [17]. Adding material in the chamber such as electrodes and macor insulators reduces the vacuum quality by *both* increasing the surface area and decreasing the effective tube diameter, and thus we must be very careful to correctly account for it all. Given all these factors, the pressure in the system is given by [17]:

$$P = SA \cdot R \cdot \left(\frac{1}{S_p} + \frac{L}{78D^3} \right). \tag{7.6}$$

While it is difficult to calculate the outgassing rate, we can assume that all materials outgas the same amount ($R_{new} = R_{old}$) and simply compare the surface area of the new apparatus to that of the old apparatus, whose pressure we know. The outgassing rates of macor and tungsten are actually similar to glass and steel, and so this is a reasonable approximation.

We can now compare the pressure in the new chamber to that in the old chamber, and this ratio is given by:

$$\frac{P_{new}}{P_{old}} = \frac{SA_{new}}{SA_{old}} \frac{\frac{1}{S_{p,new}} + \frac{L_{new}}{78D_{new}^3}}{\frac{1}{S_{p,old}} + \frac{L_{old}}{78D_{old}^3}}. \tag{7.7}$$

Estimating the effective diameter of the new tube out of the science cell D_{new} requires detailed knowledge of the macor pieces in the tube. The most important of which is the back macor piece, as shown in Fig. 7.19. This piece has undergone many revisions to remove as much material as possible, because it plugs the access of the science cell to the connecting tube. Shown in grey are the macor sleeves. We found that as a result of the assembly process the rods bow out and only need to be supported from outside. This solution makes the machining much simpler, since the sleeves are very thin and weak. The groove holds the ITO-coated plate, and the hole behind the groove is where the rods come through to contact the ITO coating on the plates.

For the old chamber, we estimate $L_{old} = 76\,mm$ and $D_{old} = 25\,mm$, while for the new chamber we estimate $L_{new} = 100\,mm$ and $D_{new} = 25\,mm$. Note that these numbers are actually similar between the two cells. Given these numbers, we can estimate P_{new}/P_{old} as a function of SA_{new}/SA_{old}. Based on the geometry of all the electrodes and macor insulators, we estimate $SA_{new}/SA_{old} \approx 4$–$5$. The pumping

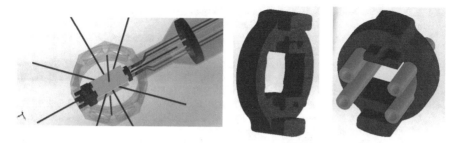

Fig. 7.19 The back macor pieces which hold the electrodes in the science cell. For vacuum conductance reasons, we want to keep this piece as small and thin as possible. Shown in grey are the macor sleeves. We found that as a result of the assembly process the rods bow out and only need to be supported from outside. This solution makes the machining much simpler, since the sleeves are very thin and weak. The groove holds the ITO-coated plate, and the hole behind the groove is where the rods come through to contact the ITO coating on the plates

speed of the ion pump in the old chamber was $S_{p,\text{old}} = 35\,L/s$, and as expected $P_{\text{new}}/P_{\text{old}}$ is minimized when $S_{p,\text{new}}$ is large. Nevertheless, $P_{\text{new}}/P_{\text{old}}$ saturates for $S_{p,\text{new}} \approx 50\,L/s$, and the value depends on $SA_{\text{new}}/SA_{\text{old}}$. $P_{\text{new}}/P_{\text{old}} = 3$ for $SA_{\text{new}}/SA_{\text{old}} = 3$, and 5 for $SA_{\text{new}}/SA_{\text{old}} = 5$. Note that if $D_{\text{new}} = 20\,$mm instead, these numbers scale up by ≈ 1.7.

Since $P_{\text{old}} \ll 10^{-11}$ mbar, it is difficult to measure with a pressure gauge, so we measure the $1/e$ lifetime of atoms in a conservative magnetic trap $\tau_{\text{old}} \approx 300$ s. Therefore, since $P \sim \tau^{-1}$ and we expect $P_{\text{new}}/P_{\text{old}} = 5$, then the lifetime in the science cell should be $\tau_{\text{new}} \approx 60$ s. I will discuss this more in Chaps. 8 and 9, but since we only use an optical dipole trap in the science cell, the lifetime is limited by light scattering and heating and thus we cannot measure the vacuum-limited lifetime. Nevertheless, we have measured $\tau_{\text{new}} > 30(5)$ s in the science cell, which is in good agreement with these rough estimates. As stated above, we perform evaporative cooling in a magnetic trap outside of the science cell. This trap also has a loss mechanism (discussed in Chap. 9), but we use it to measure the lifetime in the small cube outside the science cell to be $\tau_{\text{cube}} > 150(10)$ s. Note that we added the NEG pump which has a pumping speed of $\approx 400\,L/s$, and this was not included in the above analysis. We added this to be as safe as possible since pressure problems are notoriously time-consuming to solve.

References

1. H.L. Bethlem, G. Berden, G. Meijer, Decelerating neutral dipolar molecules. Phys. Rev. Lett. **83**, 1558–1561 (1999)
2. J.R. Bochinski, E.R. Hudson, H.J. Lewandowski, G. Meijer, J. Ye, Phase space manipulation of cold free radical OH molecules. Phys. Rev. Lett. **91**, 243001 (2003)
3. M. Born, E. Wolf, *Principles of Optics* (Cambridge University Press, Cambridge, 1999)

4. L.D. Carr, D. DeMille, R.V. Krems, J. Ye, Cold and ultracold molecules: science, technology and applications. New J. Phys. **11**(5), 055049 (2009)
5. C. Chin, R. Grimm, P. Julienne, E. Tiesinga, Feshbach resonances in ultracold gases. Rev. Mod. Phys. **82**, 1225–1286 (2010)
6. J.P. Covey, S.A. Moses, M. Garttner, A. Safavi-Naini, M.T. Miecnkowski, Z. Fu, J. Schachenmayer, P.S. Julienne, A.M. Rey, D.S. Jin, J. Ye, Doublon dynamics and polar molecule production in an optical lattice. Nat. Commun. **7**, 11279 (2016)
7. J.P. Covey, L. De Marco, K. Matsuda, W.G. Tobias, S. A. Moses, M. T. Miecnikowski, G. Valtolina, D. S. Jin, J. Ye, A new apparatus for enhanced optical and electric control of ultracold krb molecules. (2017, in preparation)
8. M.H.G. de Miranda, A. Chotia, B. Neyenhuis, D. Wang, G. Quéméner, S. Ospelkaus, J.L. Bohn, J. Ye, D.S. Jin. Controlling the quantum stereodynamics of ultracold bimolecular reactions. Nat. Phys. **7**(6), 502–507 (2011)
9. D. DeMille, Quantum computation with trapped polar molecules. Phys. Rev. Lett. **88**, 067901 (2002)
10. A. Frisch, M. Mark, K. Aikawa, S. Baier, R. Grimm, A. Petrov, S. Kotochigova, G. Quéméner, M. Lepers, O. Dulieu, F. Ferlaino, Ultracold dipolar molecules composed of strongly magnetic atoms. Phys. Rev. Lett. **115**, 203201 (2015)
11. M.W. Gempel, T. Hartmann, T.A. Schulze, K.K. Voges, A. Zenesini, S. Ospelkaus, Versatile electric fields for the manipulation of ultracold NaK molecules. New J. Phys. **18**(4), 045017 (2016)
12. A.V. Gorshkov, S.R. Manmana, G. Chen, J. Ye, E. Demler, M.D. Lukin, A.M. Rey, Tunable superfluidity and quantum magnetism with ultracold polar molecules. Phys. Rev. Lett. **107**, 115301 (2011)
13. M. Gröbner, P. Weinmann, F. Meinert, K. Lauber, E. Kirilov, H.-C. Nägerl, A new quantum gas apparatus for ultracold mixtures of K and Cs and KCs ground-state molecules. J. Mod. Opt. 1–11 (2016)
14. R. Guenther, *Modern Optics* (Wiley, New York, 1990)
15. K.R.A. Hazzard, B. Gadway, M. Foss-Feig, B. Yan, S.A. Moses, J.P. Covey, N.Y. Yao, M.D. Lukin, J. Ye, D.S. Jin, A.M. Rey, Many-body dynamics of dipolar molecules in an optical lattice. Phys. Rev. Lett. **113**, 195302 (2014)
16. K.R.A. Hazzard, S.R. Manmana, M. Foss-Feig, A.M. Rey, Far-from-equilibrium quantum magnetism with ultracold polar molecules. Phys. Rev. Lett. **110**, 075301 (2013)
17. D.J. Hucknall, A. Morris, *Vacuum Technology: Calculations in Chemistry* (Royal Society of Chemistry, Cambridge, 2003)
18. N.M. Jordan, Y.Y. Lau, D.M. French, R.M. Gilgenbach, P. Pengvanich, Electric field and electron orbits near a triple point. J. Appl. Phys. **102**(3), 033301 (2007)
19. M. Lu, N.Q. Burdick, S.H. Youn, B.L. Lev, Strongly dipolar Bose-Einstein condensate of dysprosium. Phys. Rev. Lett. **107**, 190401 (2011)
20. A. Micheli, G.K. Brennen, P. Zoller, A toolbox for lattice-spin models with polar molecules. Nat. Phys. **2**, 341–347 (2006)
21. S.A. Moses, J.P. Covey, M.T. Miecnikowski, B. Yan, B. Gadway, J. Ye, D.S. Jin, Creation of a low-entropy quantum gas of polar molecules in an optical lattice. Science **350**(6261), 659–662 (2015)
22. B. Neyenhuis, B. Yan, S.A. Moses, J.P. Covey, A. Chotia, A. Petrov, S. Kotochigova, J. Ye, D.S. Jin, Anisotropic polarizability of ultracold polar $^{40}K^{87}Rb$ molecules. Phys. Rev. Lett. **109**, 230403 (2012)
23. K.-K. Ni, S. Ospelkaus, M.H.G. de Miranda, A. Pe'er, B. Neyenhuis, J.J. Zirbel, S. Kotochigova, P.S. Julienne, D.S. Jin, J. Ye, A high phase-space-density gas of polar molecules. Science **322**(5899), 231–235 (2008)
24. J.W. Park, S.A. Will, M.W. Zwierlein, Ultracold dipolar gas of fermionic $^{23}Na^{40}K$ molecules in their absolute ground state. Phys. Rev. Lett. **114**, 205302 (2015)
25. G. Quéméner, P.S. Julienne, Ultracold molecules under control! Chem. Rev. **112**(9), 4949–5011 (2012)

26. L. Reichsöllner, A. Schindewolf, T. Takekoshi, R. Grimm, H.-C. Nägerl. Quantum engineering of a low-entropy gas of heteronuclear bosonic molecules in an optical lattice. arXiv:1607.06536v1 (2016)

27. J.F. Sherson, C. Weitenberg, M. Endres, M. Cheneau, I. Bloch, S. Kuhr, Single-atom-resolved fluorescence imaging of an atomic Mott insulator. Nature **467**, 68–72 (2010)

28. B.K. Stuhl, Ultracold molecules for the masses: evaporative cooling and magneto-optical trapping. PhD thesis, University of Colorado, Boulder (2012)

29. B.K. Stuhl, M. Yeo, M.T. Hummon, J. Ye, Electric-field-induced inelastic collisions between magnetically trapped hydroxyl radicals. Mol. Phys. **111**(12–13), 1798–1804 (2013)

30. S.V. Syzranov, M.L. Wall, V. Gurarie, A.M. Rey, Spin–orbital dynamics in a system of polar molecules. Nat. Commun. **5**, 5391 (2014)

31. S. Timoshenko, S. Woinowsky-Krieger, *Theory of Plates and Shells* (McGraw-Hill, New York, 1959)

32. S.Y.T. van de Meerakker, N. Vanhaecke, H.L. Bethlem, G. Meijer, Transverse stability in a stark decelerator. Phys. Rev. A **73**, 023401 (2006)

33. M.L. Wall, E. Bekaroglu, Lincoln D. Carr, Molecular Hubbard hamiltonian: field regimes and molecular species. Phys. Rev. A **88**, 023605 (2013)

34. M.L. Wall, K.R.A. Hazzard, A.M. Rey, Quantum magnetism with ultracold molecules, in *From Atomic to Mesoscale*, chap. 1, pp. 3–37 (World Scientific, Singapore, 2015)

35. C. Weitenberg, M Endres, J.F. Sherson, M. Cheneau, P. Schauss, T. Fukahara, I. Bloch, S. Kuhr, Single-spin addressing in an atomic Mott insulator. Nature **471**, 319–324 (2011)

36. B. Yan, S.A. Moses, B. Gadway, J.P. Covey, K.R.A. Hazzard, A.M. Rey, D.S. Jin, J. Ye, Observation of dipolar spin-exchange interactions with lattice-confined polar molecules. Nature **501**(7468), 521–525 (2013)

37. B. Zhu, G. Quéméner, A.M. Rey, M.J. Holland, Evaporative cooling of reactive polar molecules confined in a two-dimensional geometry. Phys. Rev. A **88**, 063405 (2013)

38. J.J. Zirbel, K.-K. Ni, S. Ospelkaus, J.P. D'Incao, C.E. Wieman, J. Ye, D.S. Jin, Collisional stability of fermionic Feshbach molecules. Phys. Rev. Lett. **100**, 143201 (2008)

Chapter 8
Designing, Building, and Testing the New Apparatus

With a promising set of ideas in place for the design of the new apparatus, it is time to discuss the building and testing of the new chamber, coils, and imaging systems for the new apparatus. We begin with the assembly and testing of the electrodes and their ability to reach large electric fields.

8.1 Reaching Large Electric Fields

Testing of the electrodes was done in many test configurations in small chambers pumped with turbo-molecular pumps. While these tests all had promising outcomes, the performance of the electrodes in the fully assembled new apparatus chamber is ultimately what matters. In order to do these tests, however, the full chamber (or at least the science side) must be assembled. Chapter 9 will describe in more detail how the chamber is in two halves: the science half and the MOT half. They are connected with a differential pumping tube, a flexible bellows, and a gate vacuum which allows the two vacuum regions to be separated.

8.1.1 Macor Design

To build the full electrode assembly, we must start with the design and fabrication of the macor insulating pieces. As was discussed in Chap. 7, we want the surface path length between adjacent electrodes to be $\approx 10\times$ their separation [4, 7]. Figure 8.1 shows some of the macor pieces used to hold the electrodes. The one in the right image is the front holder in the science cell, which gently presses against the window at the end of the cell. Note that its diameter is ≈ 15 mm, or roughly the size of a dime. The insulators on the left hold the electrodes from the small cube to the science cell.

© Springer Nature Switzerland AG 2018

J. P. Covey, *Enhanced Optical and Electric Manipulation of a Quantum Gas of KRb Molecules*, Springer Theses, https://doi.org/10.1007/978-3-319-98107-9_8

Fig. 8.1 The macor insulators which hold the electrodes in the science cell. The smallest piece which goes in the science cell is compared to a dime for reference

They are covered in blue glue, because the machining was performed with a slab of macor glued to a block of aluminum. This glue is one of many contaminants which must be carefully removed during the cleaning process, to be discussed later.

There are six rods that go into the science cell: the four rods between the ITO-coated plates and the two rods which contact the ITO coatings on the two plates. These six rods are arranged as a perfect hexapole down the transmission line from the feedthroughs to the science cell. The small piece in the left image is an old version of the back electrode in the science cell, which resides near the entrance to the tube that is connected to the science cell. Even this amount of material would have significantly reduced the vacuum conductance out of the science cell and therefore limited the pressure achievable in the science cell.

Note the slots on the bottom macor piece in the left image. This is the piece that holds the electrodes in the flange of the science cell. These slots are essential to indicate the entire electrode assembly in the cell relative to the windows. The relative angle between the ITO-coated plates and the cell top and bottom windows was minimized by rotating the electrode assembly in these slots, and finally fixing the orientation. The holes for the electrodes on this piece have long sleeves because of the orientation of the slots with respect to the rods, which requires one of the slots to be adjacent to one of the rods that connects to the ITO-coated plates. These rods have the largest magnitude potential, and thus it is particularly important to maximize its path length to ground.

8.1.2 Connecting the Electrodes to the ITO Coatings

The two rods on the top and bottom of the hexapole guide connect to the ITO coatings on the fused silica plates. This is done by facing off the rod through half of its 1 mm diameter, leaving a flat section 1 mm wide and typically 3–4 mm long.

These rods are made of tungsten, which is hard to cut or grind. The facing of these rods as well as any other cutting of tungsten rod described throughout was done with a water-cooled diamond-blade grinder.

The plates sit in grooves on the two macor pieces that hold all the electrodes together in the science cell, which are shown in Figs. 7.3 and 7.19. The ITO coating on the plates extends to $\approx$1 mm from the ends along the long axis, and so the part of the plate that fits into the grooves on the macor holders is uncoated. This region is required to hold the plates during their ITO/AR coating process. Figures 7.3 and 7.19 also show a hole in the back macor piece of diameter 1 mm through which the rod extends to connect to the plate. The center of this hole is lined up with the bottom surface of the plate, such that when the rod is faced off half way through, it will contact the plate. No epoxy or conductive glue is used, so once the rod is faced off, it is angled in slightly such that the connection to the plate is maintained by the resulting cantilever spring force.

To ensure that the plate is centered on the macor holders and the rod connector, a small slot is grinded $\approx$1 mm into the end of the plate with the diamond-blade grinder. The rod then fits into this slot, and the faced-off last several mm extend into the ITO-coated region and contact the coating. This approach also makes the connection between the rod and the ITO more robust. We measure Ω-level resistance between the rod and the ITO coating, which is negligibly small compared to the $\approx$1 kΩ resistance of the ITO coating itself. This connection can be seen by looking carefully at Fig. 8.15.

8.1.3 Electrode Surface Smoothness

Stark decelerators are typically designed using the least magnetic stainless steel, SS316. The smoothness of the surfaces of their electrodes is very important and is often the limiting factor on the current drawn during steady-state operation. SS316 can be electropolished nicely in a 10% NaOH solution, which we did for all the steel parts of the electrode system. However, the electrodes between the small cube and the science cell are tungsten to mitigate any magnetization problems, which cannot be easily electropolished to my knowledge. In fact, we sent some tungsten rods to be electropolished by a company which claimed to be able to electropolish tungsten, and they came back with craters and more surface roughness than they had originally. Images of electropolished SS316 and tungsten can be seen in Fig. 8.2. We expect that the challenge with tungsten is that the rod is a sintered material, which is likely the cause of the cratering upon electropolishing.

After exhausting the possibilities of electropolishing tungsten, we turned to the idea of mechanical polishing. We found that a buffing wheel and buffing compound work very well to polish tungsten. Buffing compound is an oil-based clay, called a vehicle, in which are embedded micro-abrasives. We found that a gray compound works very well as a coarse polish, and then a red compound works very well as a fine polish. This procedure was so effective that we could make our tungsten

Fig. 8.2 Electropolished
steel and tungsten rods. The
lower rod is SS316, and the
upper rod is tungsten. This
image was taken with an
optical microscope

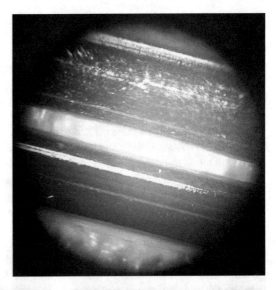

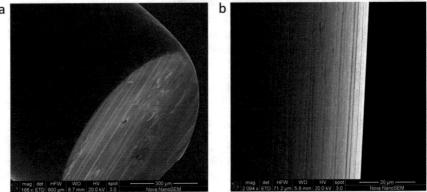

Fig. 8.3 An SEM image of mechanically polished tungsten rods. (**a**) The end of the rod, with the
scale bar indicating 300 μm, and (**b**) the side of the rod with the scale bar indicating 20 μm

rods smoother through mechanical polishing than electropolished SS316 [4]. We
measured the surface quality using a scanning electron microscope (SEM), and such
images are shown in Fig. 8.3. The long, axial ridges seen in Fig. 8.3b are expected
to be due to the mechanics of the buffing wheel, which was mostly spinning against
that axis.

Unfortunately, the mechanical polishing approach has a huge downfall of which
we did not become aware until we prepared a vacuum system and attempted to
reach ultrahigh vacuum ($<10^{-11}$ mbar). The cleaning procedure was unable to
completely remove this clay, which persisted in the vacuum system with a relatively
high outgassing rate. The outgassing was low enough that the operation with the
turbo-molecular pump during the vacuum pump-down procedure was typical, and

pressures in the 10^{-8}–10^{-9} range could easily be reached. In fact, at the same time that we were working on this, Noah Fitch in Heather Lewandowski's group was building their traveling-wave stark decelerator, which is made of tantalum electrodes. They similarly had a difficult time in electropolishing tantalum, so they decided to electropolish and mechanically polish their electrodes with the same clays that we used. Their vacuum system is very different than ours, and they were ultimately able to reach sub-10^{-10} mbar in their large, open chamber using several gigantic turbo pumps.

As discussed in Chap. 7, our vacuum system needs to reach sub-10^{-11} mbar pressures, and the pumping speeds and conductances are limited. Therefore, the additional gas load from the buffing compound is unacceptable. Indeed, we found that when we baked the chamber, the pressure was higher after the bake than before, indicating that the clay had been activated and dispersed at high temperature. The pressure levels at this stage required an ion pump to reach, and thus the dispersal of this clay got into the ion pump and rendered it unusable. It took us several tests to learn that the problem was indeed coming from the clay, and several ions pumps were contaminated in the process, since the pressure would saturate the range of a turbo pump. In the end, this was a rather expensive and frustrating process.

We were never able to find smoking gun evidence that the clay was causing the problem, merely a 100% correlation during the vacuum tests. The only way we could imagine detecting clay residue on the tungsten rods was with the SEM. Dielectric materials appear white on an SEM because they reflect electrons, unlike conductors like tungsten. Therefore, we suspected that perhaps we could find white clumps near points of elevated surface topography like bumps or ridges. Figure 8.4 shows such images for polished and unpolished rods. We certainly found more white clumps on the polished rod, but we also found some white-ish sections on the unpolished rod. Hence, I would say that this measurement was suggestive and perhaps even circumstantial evidence, but still not a definitive proof.

 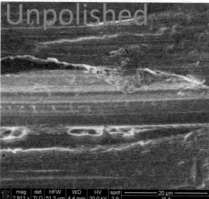

Fig. 8.4 SEM images of blemishes on the tungsten rods, with and without being mechanically polished. White objects are dielectric materials, since they reflect electrons

Since both electropolishing and mechanical polishing with buffing compound have now been ruled out, we decided to test how important it actually is that the tungsten rods are polished. Stark decelerators often reach $E \approx 100\,\text{kV/cm}$, which is well beyond the fields we wish to reach. Moreover, Stark decelerators are designed to have relatively small fields at the insulators such that they are not limited by surface currents on the insulators, as discussed in Chap. 7. However, our macor insulators are in the positions with the largest electric fields, and thus they will limit the ultimate field we can reach. Therefore, it is possible that our system would never be limited by the smoothness of the electrodes.

8.1.4 Conditioning the Electrodes

Therefore, instead of mechanical polishing with buffing compound we used three consecutively finer sandpapers of grits 2400, 4000, and 12,000. These were primarily used on the ends of the rods and the faced-off sections that contact the ITO-coated plates. We performed many tests with just two rods, both smooth and faced off at the end. We found that the limitation of rod smoothness only comes from small burrs and spikes on the faced-off rods, but the ultrafine sand paper is actually quite helpful to mitigate these problems.

Conditioning of electrodes is a violent process by which small spikes and other features on the surface of the rods ablate each other by shooting a high-energy electron beam. These features serve as nucleation sites like a lightning rod, and eject or receive an electron beam. We performed a conditioning procedure, similar to what is used in Stark decelerators [2, 4, 7]. This conditioning data was taken using molybdenum plates instead of ITO-coated glass, for reasons to be discussed in a later section. Conditioning is done by progressively increasing the voltages to the electrodes, and measuring the current response as a function of time. A schematic of the system is shown in Fig. 8.5. We performed most conditioning measurements with the same voltage on the top plate and top rods such that only two voltages were used $\pm V_{\text{test}}$.

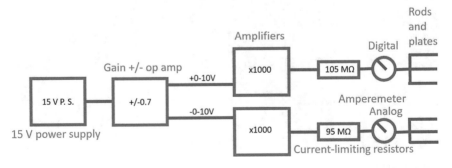

Fig. 8.5 A schematic of the conditioning layout. The 0.7 gain op amp circuit was used to ensure that the voltage to the amplifier never exceeded $\pm 10\,\text{kV}$

Fig. 8.6 The voltage and
current during a typical
conditioning measurement.
Such small spikes are
typically 5–10 μA

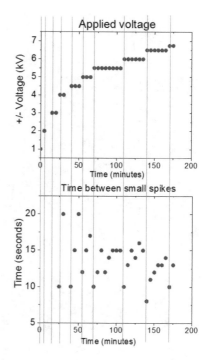

To limit the current that can be drawn during the electron beam ablation, we put
100 MΩ resistors in series with the rods, and thus $I_{lim} = V_{test}/R = 100\,\mu A$ for
$V_{test} = 10\,kV$. However, this is still a lot of current. An interesting feature of our
apparatus is that the electrodes are quite visible inside the glass cell, as opposed to a
Stark decelerator. Moreover, in a dark room the human eye is nearly a single photon
detector in the blue wavelength range, and thus we can literally see every electron
emitted by the electrodes during conditioning. Accordingly, this measurement is
both really exciting and really terrifying.

Figure 8.6 shows typical data from a conditioning measurement. The upper plot
shows $\pm V_{test}$ as a function of the time during conditioning. Note that progressively
more time is spent at each voltage, since it takes longer for the transient current
response to stabilize. As discussed above, the smoothness of the tungsten rods
(besides the faced-off connections to the plates) was never a source of large
current and never limited the voltages we could reach. Rather, the development
of small surface paths on the macor insulators in the science cell and throughout
the transmission line to the feedthroughs has always been the dominant breakdown
mechanism. The total surface area of the electrodes in our system is incredibly
small compared to a Stark decelerator. Therefore, each individual current spike,
or electron emission, could be detected. Instead of drawing a finite current that
changed slowly, we often drew very little current ≈ 1–$2\,\mu A$ and had discrete spikes
of ≈ 5–$10\,\mu A$ beyond the background, whose frequency would vary.

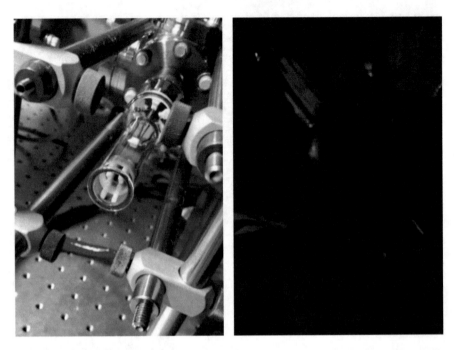

Fig. 8.7 Observation of blue flashing in the science cell during the conditioning measurement. The blue emission is localized to part of the front macor holder which is gently pressed against the science cell's end window

The lower plot shows the time in seconds between these discrete spikes. For $V_{test} < 4$ kV, these spikes were very infrequent, and spaced by >60 s. As the voltage increased, however, the spike frequency increased until it saturated at 10–15 s/spike. This is a very encouraging response, and the field reached in this test was 32 kV/cm. We typically only apply the electric field for ≈ 1 s, so it is unlikely that any damage could occur at these timescales even at larger voltages. However, it would be beneficial to understand where on the electrode system the spikes are occurring. Any spikes inside the science cell would be visible by eye, and so we can carefully look for flashing during these conditioning measurements.

Figure 8.7 shows the observation of some blue flashing which was observed inside the Pyrex test cell that was used for these particular conditioning measurements. The location of the emission is on the left edge of the front macor holder which is gently pressed against the science cell's end window. Thus, the breakdown mechanism involves the macor rather than the surface smoothness, as anticipated. While conditioning is a self-healing process that smoothes the electrodes, the development of surface currents in the insulators is a run-away problem that will continue to lower the voltages that can be reached in the experiment. Indeed, during the conditioning of a Stark decelerator it is common to go to voltages a few kV above the operating voltages such that operation is even more stable. However, when

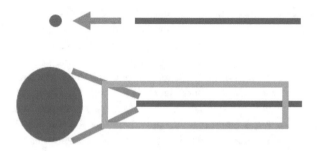

Fig. 8.8 A cartoon on the observation of triple-point-induced electron beam breakdown. The circles on the left are macor powder dust on the end window, and the orange lines or arrows show the electron beam. The presence of the macor insulator (green) beyond the electrode (blue rod) has an enormous bearing on our observations

destructive processes occur in the insulators this is not an option. Therefore, it is important that we do not consistently operate at such voltages where this breakdown occurs in the final assembly.

Another important limitation of having the insulators in the region of maximal electric field is the presence of triple points, as discussed in Chap. 7, which can lead to rf breakdown and the development of an electron beam. These issues are particularly acute in a small glass cell where other dielectric surfaces are relatively near the triple points. The front macor holder used in this test had macor sleeves which extend well beyond the end of the rod electrodes, thus leading to θ being small as discussed in Chap. 7. The generation of an electron beam was clearly apparent during some of our measurements at high voltages. This beam only originated from the rods with negative potentials, as expected, and the beam was clearly directed towards the end window.

The evidence of this beam was the development of a circular deposit of macor powder on the end window. The generation of this instability was quasi-periodic, where every several seconds a small burst of macor powder would hit the end window followed by a large burst. We believe that this timescale corresponds to the growth of an rf breakdown instability near the triple point. Figure 8.8 shows a cartoon of what we observed. The circles on the left are macor powder dust on the end window, and the orange lines or arrows show the electron beam. The presence of the macor insulator (green) beyond the electrode (blue rod) has an enormous bearing on our observations. It appears that the solid angle subtended by the electron beam is influenced by the presence of the macor, and the macor dust depositing on the end window is caused by the electron beam penetrating the macor sleeve. Note that one of the negative rods had this problem while the other did not, which suggests that it could be exacerbated by a particular rounding shape of the end of the rod. Nevertheless, this problem has been entirely solved simply by having the macor sleeve not reach the end of the rod, as discussed in Chap. 7.

While these macor-related problems must be taken seriously, there is a lot of reason to be optimistic. First, the triple-point issue has a simple solution which immediately corrected the problem. Second, as discussed in Chap. 7 the homogeneous electric field occurs when the rod voltages are at 42.25% of the plate voltages, and thus will never exceed ± 4.225 kV. The surface currents on the front macor insulator developed at $\approx \pm 7$ kV, and thus are negligible in the homogeneous field configuration. Third, we operate these electric fields with a very small duty cycle. Unlike Stark decelerators that use large electric fields to produce cold molecules, we operate an ultracold atom experiment where molecular manipulation with large electric fields only happens for roughly 1 s every 30 s. Fourth, the onset of the macor-related instabilities takes several seconds. Therefore, even if they were limitations, they could potentially be circumvented simply by applying the field for short times. Fifth, as discussed in Chap. 7, J_z is actually maximized at the relatively low field of 10 kV/cm, and so such large fields are necessary only for a limited number of experiments, such as evaporative cooling of molecules.

8.1.5 Reaching Large Fields with the ITO-Coated Plates

The above conditioning measurements were done with molybdenum plates rather than ITO-coated fused silica plates, and thus we must repeat some of these measurements using ITO-coated glass plates instead. One particular test we did was just with two ITO-coated plates, and the faced-off rods that connect to them as discussed above. This test was done without using the ultrafine sandpaper mentioned above, and elucidates the need for such a procedure. The faced-off rods have sharp edges and small burrs which ablate away throughout the conditioning process as expected. The proximity of the ITO-coated plates, however, makes this ablation process particularly damaging. The negative faced-off rod shoots electrons at the positive faced-off rod, but they subtend some angle around the faced-off rod and hit the plate. Figure 8.9 shows the damage to the ITO coatings that result, and it is clearly much worse on the plate which had the positive voltage. The edge of the ITO coating and connection slot are apparent at the bottom of the plates.

The dark moat on the positive plate around the contact point of the rod shows how the coating was locally destroyed, and the ITO at the contact point is completely electrically isolated from the rest of the coating. This is obviously a huge problem as the ability to control the plate voltages is lost and they simply float to whatever potential. Moreover, there is some evidence of damage in one of the corners at the far end of the plates. This is due to the makeshift way the plates were held where there was no support at the far end, and the plates got too close together. Nevertheless, these measurements show the propensity for damage from such fragile coatings in such a violent atmosphere.

Our solutions to this problem are two-fold. First, the ultrafine sandpaper enormously improved the smoothness of the faced-off sections of the rods and mitigated the ablation when applying large fields. However, it is still not quite smooth enough to entirely avoid damaging the ITO-coating. Second, the final assembly was done

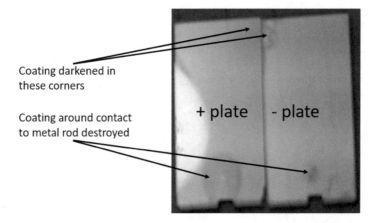

Coating darkened in
these corners

Coating around contact
to metal rod destroyed

+ plate   - plate

Fig. 8.9 Damaged ITO-coated plates after conditioning. Notice that the damage on the positive plate is much worse than the negative plate

first without any plates present, and the conditioning was done as such. Then, once the conditioning had been performed and the faced-off rods were exquisitely smooth, the vacuum system was opened and the ITO-plates were inserted. This approach allowed us to reach fields of ≈40 kV/cm without the plates with minimal current draw and no discharge or spiking in the science cell, and then ≈30 kV/cm when the plates were added before the onset of any spiking in the science cell.

8.1.6   Large Fields with High Intensity Beams Through the ITO

I will now discuss additional conditioning measurements using ITO-coated plates where high intensity beams are incident on the coating during the conditioning measurement. It is possible that the photons from a high intensity beam could facilitate the emission of electrons from the conductive coating, in accordance with the photo-electric effect. Fortunately, the work function of ITO is ≈4.2–4.8 eV depending on the coating method and thickness, which is quite high. Nevertheless, the energy of the work function corresponds to an optical photon of wavelength of 280 nm, within a factor of 5–6 of our lattice wavelength of $\lambda = 1064$ nm.

We performed the conditioning measurements with a high intensity beam focused on the ITO-coatings, as shown in Fig. 8.10. The highest intensity we anticipate using in the experiment is a 5 W beam focused to a $1/e^2$ radius of 50 μm. Therefore, to match this intensity we used a 1 W beam focused to 20 μm. The Rayleigh length of such a small beam is sub-mm, however, and so we used a translating mount for the focusing lens to ensure that we could have the highest intensity incident on the ITO-coatings. During the conditioning measurements, there was no effect of the focused beam on the current drawn, and no discharging was

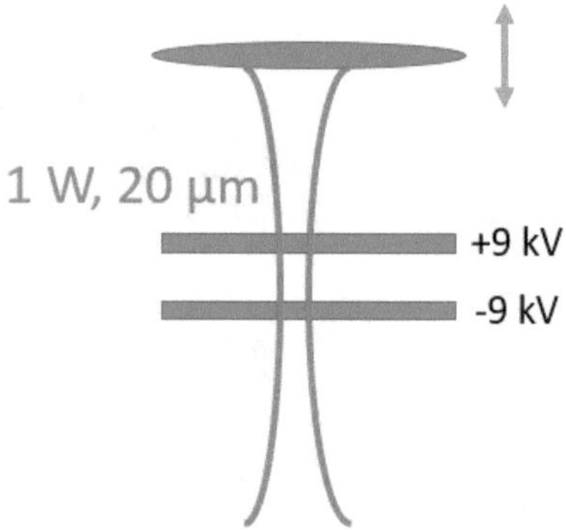

Fig. 8.10 The setup for ITO-coated plate conditioning with an incident high intensity beam. The focusing lens could be translated to ensure that the highest intensity was hitting the ITO coating

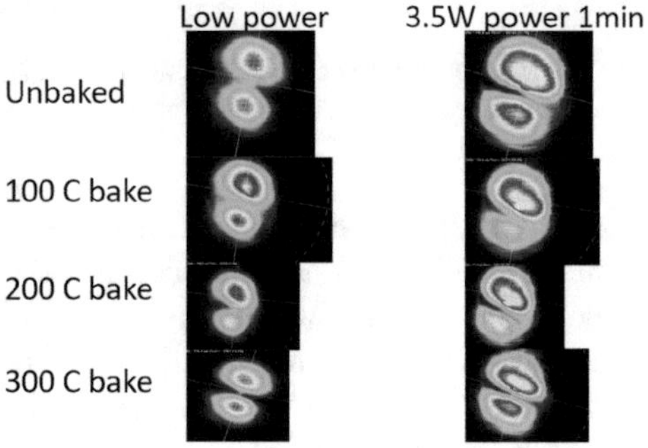

1 W, 20 μm

+9 kV

-9 kV

Fig. 8.11 The output mode of a Mach–Zehnder interferometer with an ITO-coated plate in one arm. This mode was measured for low power and high power for several conditions of the ITO-coated plate, such as before and after baking to a high temperature

Low power   3.5W power 1min

Unbaked

100 C bake

200 C bake

300 C bake

observed even up to potentials of ±9 kV, or 30 kV/cm. This was an excellent result for the operation of large electric fields with an incident high intensity beam.

Another question, however, is what effect the high intensity has on the optical properties of the ITO-coating. Since ITO is a semiconducting material, it has finite absorption near the band gap in the IR range, and the absorption at $\lambda = 1064$ nm is proportional to the coating thickness. While an AR coating paired with the ITO can alter the reflection and transmission properties defined by $\mathrm{Re}(n(\lambda))$ where n is the index of refraction, the absorption caused by $\mathrm{Im}(n(\lambda))$ is an unalterable property of the material. As discussed later, the absorption of our ITO coatings is $\approx 2\%$, which could cause heating that in turn changes the transmission or absorption, or even causes thermal lensing.

To study this, we put an ITO-coated plate in one arm of a Mach–Zehnder (MZ) interferometer. The output fringe on the optical mode of the recombined beam was then monitored on a beam profile camera, as shown in Fig. 8.11. We looked at this mode for both low intensity and high intensity under several circumstances. As discussed later, we have observed that optical properties of the ITO coating change irreversibly upon baking, and so we tested an unbaked plate as well as plates that had been baked to various temperatures. For all of these cases, however, we saw no discernible change in the optical mode of the MZ interferometer output even after minutes of hold time.

8.1.7   Effect of Titanium-Sublimation Pumps on Electrode Insulators

Another potential compatibility issue in this apparatus is high voltage and ultrahigh vacuum. In Chap. 7, I discussed the vacuum conductance and outgassing limitation of the electrodes and insulators in the vacuum, and I will come back to this issue in the next section. However, we must consider the effects of the titanium-sublimation pump, which operates by sputtering titanium onto the chamber walls. Titanium is a gettering material, which means that it serves as a pump and often can be used to reach pumping speeds of $\approx$1000 L/s when sputtered onto a large enough surface area.

This type of pumping technology is rarely used in a system with nearby high voltage electrodes, and an obvious issue is whether a titanium layer could develop on any of the macor insulators. Even a nm-thick layer would cause enormous problems with surface currents developing between the electrodes. Therefore, we must place the Ti-sub pump in a position which has minimal contact with any electrode or insulator. The titanium is sputtered onto an area determined by a line-of-sight from the Ti-sub pump filaments. Therefore, we placed the pump in its own arm on a tee behind the large cube which houses the feedthroughs. The pump is activated by heating the filaments to a temperature at which titanium has a large vapor pressure and is red hot. Figure 8.12 shows a view from the small cube back into the large cube where the Ti-sub pump is behind the electrodes in an arm that is oriented vertically. The blackbody radiation from the filaments can be seen with the room lights on or off.

Fig. 8.12 Activating the Titanium-sublimation pump, as seen from the small cube. These images are from the small cube where the electrodes are turning the corner and heading to the science cell, which is to the left. The blackbody radiation from the filaments can be seen with the room light on or off

To test the effect on the electrodes of Ti sputtering in the chamber, we compared a conditioning test before and after the first activation of the Ti-sub filaments. Fortunately, we did not observe any difference in the current drawn or the nature of any current spikes. Therefore, we sufficiently shielded the Ti-sub filaments from any electrodes in the large cube area. Yet, its pumping effects are still apparent as we have been able to reach excellent pressures in this chamber ($<10^{-11}$ Torr), to be discussed later in this chapter.

8.2 Electrode Parallelism

Now, I describe the design and testing of all the electrodes, and the procedure used to assemble everything between the feedthroughs and the science cell. As was discussed in Chap. 7, the homogeneity of the field that can ultimately be reached is limited by how parallel all the plate and rod electrodes are in the science cell. I will begin by discussing how the electrodes in the science cell were assembled and what angle we were able to reach.

8.2.1 Assembling the Electrodes in the Science Cell

The entire electrode assembly inside the science cell was assembled on its own, and then inserted into the cell like a ship in a bottle. Therefore, the electrode assembly had to be quite rigid and robust, and the cell was only used to indicate the assembly relative to the chamber. I will begin by discussing how this was accomplished.

As the hexapole configuration of the rods approaches the science cell, all six rods are angled together to reach a smaller separation for larger electric fields. However, the ability to do this required that all six rods were bent in exactly the same way, and so a bending rig was needed to do this accurately. Another rig was used to taper the rods from the large cube down to the small cube, to be discussed later. These rigs are shown in Fig. 8.13 and are simply dowel pins protruding from a phenolic block. The positions of the pins were chosen to move the rod by the desired amount, and the angle was chosen based on the chamber or cell tapering around the rods.

Figure 8.14 shows the six tungsten rods that are bent to go into the science cell. The top four rods go between the ITO-coated plates, and the bottom two will be faced off to contact the plates. Note that the angle with which the bottom two rods are faced off must exactly match the plane of the bend of the rod. These six rods are rounded at the tip using a polishing wheel. This image was before any sanding or cleaning of the rods.

While tungsten is very hard, it is also a brittle material. Roughly, one third of the times we tried to bend the tungsten rods in this way to go into the science cell, they splintered at the bend like a piece of wood. The fracture was always obvious, but we looked at all the rods using an optical microscope to ensure that

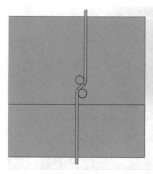

Fig. 8.13 Rigs for accurately bending the electrodes. The left image shows a solid model of the rig used to bend the electrodes near the science cell, and the right image shows both rigs with a bent rod between the dowel pins

Fig. 8.14 Tungsten rods bent to go into the science cell. The top four rods go between the ITO-coated plates, and the bottom two will be faced off to contact the plates. This image was taken after the rod tips were rounded, but before any cleaning

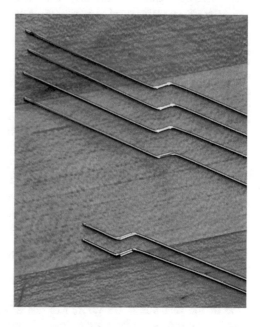

no microfractures were present. We were unable to find a correlation between the fracturing and anything else about the rods, but we assume that it must be related to how the internal domains and microlayers are aligned relative to the bend angle.

Once the rods were bent, a fine angling procedure using graphing paper was employed to ensure that the sections of the rod on either side of the bend were parallel. Moreover, the rods were placed on a flat surface such as a super-polished granite block to ensure that they lay on a plane, and that there were no angles perpendicular to the plane defined by the bend. This procedure was useful, but a final angle adjustment was done with the electrodes in their assembly.

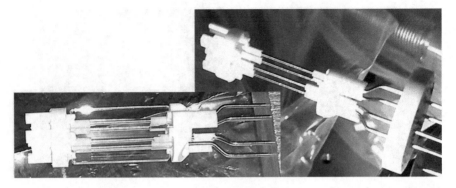

Fig. 8.15 The first assembly of the electrodes including the ITO-coated plates. The left image is with the ITO-coated plates, and the macor piece to the right of the bend is replaced with an aluminum mounting block with six holes arranged in a hexapole pattern. Note how the faced-off rod is contacted with the ITO-coated plate. The right image shows the assembly using all the macor pieces, but the plates have not yet been included

With the ability to bend the rods, we were able to assemble the electrodes to be inserted into the science cell. Figure 8.15 shows the first assembly of all the electrodes together, including the ITO-coated plates. The left image shows a metal rig just after the rods bend up to a larger radius which was used occasionally instead of the macor piece so that small tweaks to the rod angles could be made without inducing too much stress in the fragile macor pieces. The length of the rods was ultimately chosen such that the connection in the cube before the science cell gently pushed the electrode assembly against the end window of the science cell. This will be discussed more in a later section.

8.2.2   *Measuring the Parallelism and Straightness of the Rods and Plates*

One of the first things we wanted to test after assembling the electrodes is how parallel the rods and plates are to one another. The angle between the plates could be measured interferometrically, but this approach would be difficult to extend to the rods. Therefore, we used a dial indicator on a super-polished granite block to measure these angles, as shown in Fig. 8.16. The indicator was swept along the rods and plates, and the change of height was measured over their length. This allowed us to measure angles as well as bowing.

We found that the macor pieces which hold the electrodes inside the science cell had to exactly match the diameter of the rods and the thickness of the plates to minimize the angles. Any tolerance caused a much larger angle. While this may be obvious, it made the assembly much more difficult because the variation in diameter between the different plates and rods was important. Therefore, we had to choose four rods that had the same exact diameter and similarly two plates with identical

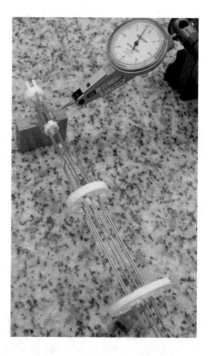

Fig. 8.16 Measuring the angles between the rod and plate electrodes. A dial indicator on a super-polished granite block was swept along the length of the rods and plates to measure their relative angles

thickness, at the level of $1/10{,}000$ inch or $2.54\,\mu m$. This was also challenging on the machining side, since the rods and plates are already very small, and macor is incredibly fragile and brittle.

Nevertheless, in doing so we were able to get the two plates parallel to within $\approx 5\,\mu m$ relative separation over their length. We were able to reach similar levels with the four rods, though slight bowing due to stress was observed. However, when we put everything together with the rods contacting the plates we found that these numbers went up to $\approx 10\,\mu m$ for the rods and plates. This is still quite the accomplishment, and puts us within the flatness goals discussed in Chap. 7, potentially allowing for investigations of quantum magnetism in a large DC field. It is important to note, however, that we do not know how much this was altered by the baking procedure of the chamber, and the moving of the new chamber (see Chap. 9). Since tungsten and macor both have very low coefficients of thermal expansion and tungsten is very hard, we expect that these numbers have changed only slightly.

8.3 AC Electric Field Coupling

Now, I will go on to discuss the assembly of the electrodes from the feedthroughs to the large cube. This region includes the coupling capacitors and decoupling inductors in the bias tee. This discussion will include measurements of the AC coupling at various parts of the transmission line from the feedthrough to the science cell.

8.3.1 Building the Coupling Capacitors

The design of the coupling capacitors was described in Chap. 7, so I will now discuss how they were built. Figures 8.17 and 8.18 show the four capacitor pillars mounted on the bottom flange in the large cube. A four-pin low-power feedthrough is below this flange, and the four macor standoffs from the flange contact the four pins on the vacuum side of the feedthrough, emanating from the hole shown in the middle (see Fig. 8.18). The top of the standoffs then go to steel disks, on which the macor piece sits. This macor piece serves as the dielectric material between the capacitor plates. These steel disks serve as the bottom capacitor plates, and the top capacitor plates are the four pillars shown. The holes near the top of the plates are where the inductors from the high voltage feedthroughs will connect, and ultimately go down the transmission line to the science cell.

Even though the four-pin feedthrough is not designed for coupling AC and its impedance is nowhere near 50 Ω, we chose this design so that we could control outside the vacuum chamber how the AC signal is connected to the four rods. This will ultimately allow us to control the polarization of the microwave field inside the electrodes, and thus the angular momentum of the rotational transitions we can drive. As discussed in Chap. 7, the capacitor area and gap were chosen such that this capacitance is $\approx 10\times$ larger than the capacitance between the rods in the science cell to make an effective AC voltage divider that will optimally couple the majority of the AC power onto the rods in the science cell.

Figure 8.19 shows how the pillars for the capacitors are connected with the inductors (right) and into the transmission line which goes to the science cell (up) in the large cube. These images show a time sequence left to right of each successive

Fig. 8.17 The capacitor design and construction for coupling AC signals onto the rods. The left image shows the capacitor design on the bottom flange of the large cube, attached to which are low-power feedthrough to couple in the AC from outside the chamber. The right image shows the capacitor assembly inside the large cube

Fig. 8.18 Designing the coupling capacitors

Fig. 8.19 Assembly of the capacitors and inductors and their connection to each other and the transmission line to the science cell. These images show a time sequence left to right of each successive electrode being connected

electrode being connected. To the left is a two-pin high voltage feedthrough (Kurt J Lesker EFT1223093, Copper, rated to 30 A and 12 kV). These feedthroughs ultimately connect with the ITO-coated plates and do not go through an inductor or connect to a capacitor, because no AC is applied to the plates. To the right is a four-pin high voltage feedthrough (otherwise identical), and an inductor is connected

Fig. 8.20 The fully assembled large cube with all the inductors and capacitors connected. The top flange in the left image is where an inline valve will be placed and used for connecting the turbo pump doing vacuum preparation. The right image is from the location of the ion pump. The vertical flange on this tee is where the arm for the Ti-sub is housed

to each. These inductors connect to the pillars which are the top capacitor plates. Another rod then goes up to the transmission line to the science cell. Needless to say, the large cube assembly was very challenging to build.

Figure 8.20 shows other views of the fully assembled large cube with all the inductors and capacitors connected. The top flange in the left image is where an inline valve will be placed and used for connecting the turbo pump during vacuum preparation. The right image is from the location of the ion pump. The vertical flange on this tee is where the arm for the Ti-sub is housed. Past the macor insulating piece in the right image, the rods go to the small cube and then turn right to go to the science cell.

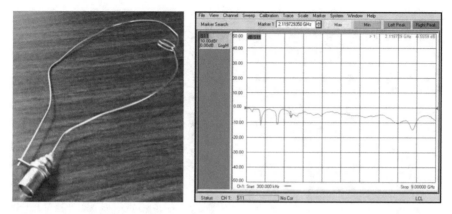

Fig. 8.21 A decoupling inductor and its resonances measured by its reflection as seen with a network analyzer

8.3.2 Designing the Decoupling Inductors

The design constraints for the decoupling inductors are less stringent since the long high voltage cables to the high voltage amplifiers are very poor AC transmission lines and are well shielded with braided grounding wires. The goal is to avoid having a resonance in the coil over the frequency range of interest, which is between 2.2 GHz and 8 GHz corresponding to the $N = 0$ to $N = 1$ transition between $0\,kV/cm$ and $30\,kV/cm$, respectively.

Figure 8.21 shows a typical coil, which is made by winding 1 mm diameter steel wire around a large screw. Various combinations of screw diameter, pitch, and number of turns were explored. The best inductor we observed is shown here. Its reflection as a function of frequency was measured using a network analyzer which goes up to 9 GHz. A large dip in the reflection shows a coil resonance where transmission is maximized, which we want to avoid. This scan shows that there are no dips between 2 and 8 GHz. The marker is at 2.1 GHz, and the two small dips to the left are below our frequency range. Most of the other inductors we tried had dips of $\approx -20\,dB$ between 3 and 6 GHz.

8.3.3 Measuring the Coupling into the Vacuum Chamber

We couple AC signals into the large cube through the four-pin feedthrough whose pins connect to the four bottom capacitor plates inside the large cube. We use a custom, yet very simple, printed circuit board (PCB) below the large cube which takes an SMA bulkhead and breaks it into four holes with the spacing of the four feedthroughs. This PCB and its connection to the feedthroughs are shown in Fig. 8.22. Two connections are AC ground, and two are AC signal. By simply

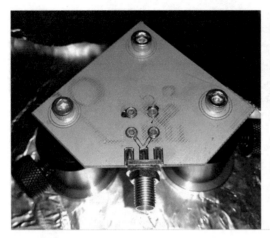

Fig. 8.22 A PCB designed to couple the AC signal into the transmission line. The left image shows the PCB whose four holes go around the four feedthroughs, and the right image shows how it all looks when connected together

changing the orientation of this PCB with respect to the feedthroughs, we can control whether the AC electric field oscillates in the horizontal plane or vertical plane in the science cell.

The first thing we wanted to measure is how effectively the AC signal is coupling onto the feedthroughs and through the capacitors in the large cube. These measurements were made with the 9 GHz network analyzer and a fast oscilloscope probe which operates up to 10's of GHz and can be touched to the electrodes with minimal perturbation to the boundary conditions of the system. We found that the insertion loss in going from the 50 Ω coaxial SMA bulkhead to the four-pin feedthroughs is $\approx$5 dB. The coupling loss through the capacitor depends on the frequency but is typically 10–20 dB. The left and middle images of Fig. 8.23 show in blue the transmission measured before and after the capacitors, respectively. These images show that low frequencies around 2–3 GHz couple very well through the capacitors, but the coupling falls off at around 6–7 GHz.

8.3.4 Measuring the Coupling Along the Transmission Line

The transmission from the top of the capacitors all the way to the science cell can be seen by comparing the middle and right images in Fig. 8.23. This was tested by glass-blowing a small hole in the Pyrex text cell shown in Fig. 8.7, and placing the AC scope probe right against the pair of rods on one side. The PCB is oriented to drive the AC field in the vertical direction, but we confirmed that the results are virtually the same when it is being driven in the horizontal direction. The

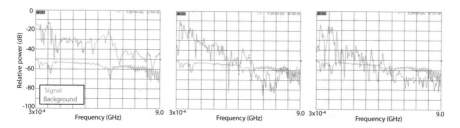

Fig. 8.23 The AC coupling through the transmission line. The left image shows the transmission (blue) through the feedthrough rods, but before the capacitors. The middle image shows the transmission (blue) through the capacitors, and on the top of the steel pillars. The right image shows the transmission (blue) to the electrodes in the science cell

transmission is generally very good, with a relative dip of at most 10 dB in at certain frequencies. This is very promising, and suggests that the AC voltage divider is working very well.

In total, we lose $\approx$25 dB on average in going from the coaxial SMA bulkhead all the way to the science cell. Our microwave amplifier allows us to put $\approx$35 dBm onto the SMA bulkhead, and so we should be able to get $\approx$10 dBm to the electrodes in the science cell. In the old apparatus, we used a large broadband horn to couple microwaves to the molecules, and the horn's transmission was severely blocked and perturbed by the IP coils around the old science cell. The polarization from the horn is roughly linear, but was completely scrambled upon going through all the coils. We used to use $\approx$10 dBm to drive the $|0, 0\rangle - |1, \pm1\rangle$ transition and $\approx$33 dBm for $|0, 0\rangle - |1, 0\rangle$.

However, it is difficult to estimate the microwave intensity within the volume of the cloud. Since the IP coils severely altered the boundary conditions for the field, it is likely that the coupling into the coils was attenuated by >10 dB. Moreover, the power from the coils subtends a large solid angle, and so the power within the volume of the cloud is also likely attenuated by a similar factor, >10 dB. The electrodes in the science cell, on the other hand, are spaced by less than a wavelength at these microwave frequencies, and thus the entire power of the microwave field that makes it that far down the transmission line will couple directly to the molecules. With these considerations in mind, we anticipate that we will be able to drive rotational transitions with Rabi frequencies similar to if not better than in the old apparatus.

8.4 Chamber Assembly Procedure

I have discussed how the electrodes are assembled within the science cell and within the large cube. It is now time to discuss the sections in between, such as going around the corner in the small cube and the reducer between the small and large cubes.

8.4.1   Turning the Corner in the Small Cube

Figure 8.24 shows the small cube and how the electrodes are connected inside. In the left image, the science cell is above. In the middle image, the science cell is to the left and the feedthroughs are to the right. The right image is taken from the MOT side of the chamber, and cold atoms come along this direction first in a magnetic quadrupole trap and then into the science cell in an optical dipole trap. As discussed above, all the electrodes are already assembled inside both the science cell and the large cube.

This small cube serves several purposes: 1) it connects the science cell electrodes to the large cube electrodes, 2) it allows cold atoms to pass in from the MOT side of the chamber such that the quadrupole magnetic trap can move over the small cube, and 3) it allows us to change from the tungsten in the science cell to the SS316 used in the large cube. This is shown in Figs. 8.24 and 8.25. The connection was made using right-angle pieces of SS316, with three different lengths on either side of the angle (see the middle image of Fig. 8.24). These right-angle pieces are connected to the rods on the science cell side and the large cube side using pieces of SS316 hypodermic tubing which is slid over both rods to be joined, and then crimped on both sides of the junction. Such pieces of tubing are shown over the rods in the left image of Fig. 8.24.

This procedure was highly nontrivial, and several custom tools were needed to successfully crimp the rods in place. The SS316 rods on the large cube side are rigidly secure. However, as discussed above, the rod positions in the science cell are not rigidly connected to anything. Thus, the goal is to gently push the two tungsten rods that connect to the plates, which will in turn move the entire electrode assembly such that the front macor piece is gently contacting the end window. Then, the other

Fig. 8.24 Connecting the electrode transmission line in the small cube. In the left image, the science cell is above. In the middle image, the science cell is to the left and the feedthroughs are to the right. The right image is taken from the MOT side of the chamber, and cold atoms come along this direction first in a magnetic quadrupole trap and then into the science cell in an optical dipole trap

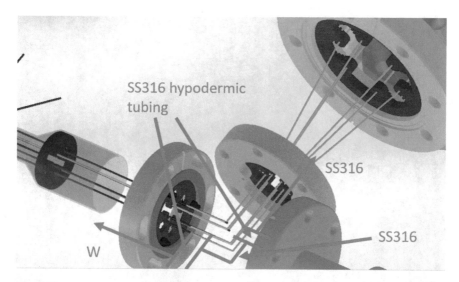

Fig. 8.25 A solid model of the electrode transition which occurs in the small cube. This figure shows how the connections are done, and how we go from tungsten in the science cell to SS316 in the large cube

four rods can be positioned such that they all extend the same amount past the macor sleeve on the front macor piece. The role of the hypodermic tubing and the right-angle pieces is to secure all six of these electrodes in place. This worked very well, but again was not easy to do.

8.4.2   The Reducer Going from the Large Cube to the Small Cube

The remaining connection to discuss is shown in Figs. 8.25 and 8.26 and is the reducer between the large cube and the small cube. Between the large cube and the reducer, and between the reducer and the small cube are double-sided blank flanges which were modified to hold the macor holders on either side of the reducer. The rod sections between these two macor holders were made with the larger rod-bending rig discussed above. The length of these rods was chosen to extend a specific distance on either side of the reducer.

On both sides, hypodermic tubing was used to connect it to the next rod section: the right-angle piece on the small cube side, and a rod to the top of the capacitor pillars on the large cube side. The large side of the reducer is shown in the right image of Fig. 8.26, and this shows how the hexapole guide in the large cube is reduced down to the hexapole guide in the small cube. Once this reducer was assembled, it was first connected to the large cube. Then, the small cube and

Fig. 8.26 The reducer between the large cube and the small cube

finally the science cell were connected to it. The NEG pump discussed in Chap. 7 is connected to the small cube opposite to the reducer.

8.5   Aligning the Electrode Assembly to the Cell Windows

Another important question arising during the assembly of the small cube part of the transmission line is how small the angle can be between the ITO-coated plates and the top and bottom windows of the glass cell. This will affect the amount of coma aberrations in the high-resolution imaging system, and will ultimately inform the resolution which we can reach without correction optics.

During the connection phase with the right-angle pieces and the hypodermic tubing, the slots in the macor insulating piece on the flange of the science cell were left loose. This allowed the electrode assembly inside the science cell to be rotated as the connections were made. The angle between the ITO-coated plate and the cell windows was measured using a beam of $\lambda = 1064$ propagating through the cell in the vertical direction. This allowed us to look at the reflections from both substrates over a long distance from the cell, from which we could measure the relative angle.

As the electrode assembly was rotated in the cell, this angle inevitably crossed zero degrees; however, it was very difficult to prevent it from rotating as the connections were made and the fasteners locked the slotted macor insulator in place. In the end, we were able to reach $<0.1°$. The coma caused by this angle is sufficiently small to allow nearly diffraction-limited performance with an NA = 0.53, but will likely become more of an issue as we try to reach NA = 0.65 in the future.

Another point of concern, however, is how much this relative angle may have changed during the extensive baking phase and the final moving of the chamber. Recently, we measured this angle again to be $\approx 0.2°$. This is starting to become more of an issue for our NA = 0.53 imaging system, but it is still sufficiently small to be relatively unimportant at this stage in the development of the experimental capabilities.

8.6 Reaching Ultrahigh Vacuum

There were many challenges that needed to be overcome before we ultimately reached excellent pressures in the new apparatus chamber. Not least of which was the contamination issues with the buffing compound. Therefore, it was imperative that none of the vacuum components we were using even contacted this compound. However, there were many other issues which were essential to reach ultrahigh vacuum.

8.6.1 Eliminating Virtual Leaks

Virtual leaks inside a vacuum system are a very well-known problem, and so it was relatively straightforward for us to avoid them sufficiently well. Of course, every vacuum chamber has virtual leaks, so the goal is really to make them small enough to be insignificant at the desired pressure. There are four main sources of virtual leaks in this system with trapped volumes that need to be vented: 1) inside the hypodermic tubing sleeves, 2) inside the holes tapped into the custom flanges which hold the macor insulating pieces, 3) inside the macor standoff below the bottom capacitor plates, and 4) inside the copper rods of the high voltage feedthroughs in which the inductors are inserted.

These four sources of trapped volume are vented with small holes in their sides. The hypodermic tubing sections were cut to length with a dremel, which was then used to make a small slit in the side towards the middle of the length. Then, the hypodermic tubing sections were mounted such that the slit was also within the trapped volume between the two adjacent rods (typically only a few mm apart). The custom flanges were tapped through in all cases where they mate with the macor insulating pieces using a fastener. This approach allows us to avoid a trapped volume at the bottom of the screws. The macor standoffs below the bottom capacitor plates have a hole in their sides so that there is no trapped volume. The same is true for the copper rods on the feedthroughs. These were nontrivial to drill, especially using isopropanol as a cutting fluid since vacuum feedthroughs cannot be cleaned easily.

There are several remaining sources of trapped gas, such as in every thread. Such sources can be quite long since their length is the integrated helix of the thread pitch, and they empty themselves very slowly, since their width is only the slop between the mating threads. This type of leak is ubiquitous, however, and seems largely insignificant even at the level of $<10^{-11}$ mbar. The different coefficients of thermal expansion between the mating materials (i.e., steel and macor) may help during the baking, since the threads of the two materials move relative to each other.

The other major source of virtual leaks in this chamber is between adjacent surfaces, such as the macor holders and the rods within the sleeves, or between the macor holders and the custom steel flanges to which they are connected. Similarly, there are large surface areas between the disks of the capacitor plates and the macor piece which serves as the dielectric material. In most of these places, additional sleeve length of macor is used to maximize the surface path length between electrodes, which again demonstrates the competing nature of high voltage and high vacuum. While these trapped volumes have been insignificant, it is likely that the baking process has helped significantly to slightly rub these surfaces against each other and open small gaps from which the trapped volumes can escape.

8.6.2 Cleaning Procedures

Due to the large number of surfaces and materials in the chamber and the vacuum requirements of $<10^{-11}$ mbar, we developed a specialized procedure for preparing the vacuum chamber. Some of these steps originated from our unsuccessful efforts to remove the buffing compound from the electrodes, and other steps were helpful to breakdown the wax or paint used in the machining procedure for macor. It is not clear how important all of these steps are when there is no buffing compound in use. For example, dichloromethane is excellent at removing tough greases and even dissolves clays.

I will begin by listing the procedure for the metal parts in the chamber, which includes all the chamber components, screws for electrode-related assembly, and all the electrodes:

- 10 min in the ultrasonic with Bio T Max citrus solution (highest setting)
- Rinse with water and Simple Green degreasing detergent (all-purpose cleaner)
- 10-min soak in dichloromethane (just a rinse for chamber parts too large to soak)
- 10 min in the ultrasonic in Alconox detergent and water (highest setting)
- Rinse in deionized water
- 10 min in the ultrasonic with acetone (highest setting)
- 10 min in the ultrasonic with isopropanol (highest setting)
- 10 min in the ultrasonic with methanol (highest setting)
- 2–3 h air bake at 250–300 °C for the chamber parts (electrodes not baked to avoid oxidation, which would affect the conditioning and electron discharging)

The baking procedure for the macor pieces is similar but a few of the steps are different. Since macor is very fragile, we used the ultrasonic on its lowest setting. The macor procedure is:

- Soak in Bio T citrus solution
- Rinse with water and Simple Green degreasing detergent (all-purpose cleaner)
- 10-min soak in dichloromethane
- Air bake overnight at 600–700 °C
- 10 min in the ultrasonic in Alconox detergent and water (lowest setting)
- Rinse in deionized water
- 10 min in the ultrasonic with acetone (lowest setting)
- 10 min in the ultrasonic with isopropanol (lowest setting)
- 10 min in the ultrasonic with methanol (lowest setting)
- Air bake overnight at 600–700 °C

Note that for all the ultrasonic sequences with macor and with the metals, the heater in the machine was on and temperature of the solution reached $\approx$70 °C. At these temperatures, acetone, isopropanol, and methanol had very high vapor pressures and were even on the verge of boiling.

8.6.3 Chamber Baking and Pump Activation Procedure

The procedure for baking the chamber and activating all the pumps is also quite complicated, partially because of the complexity of the electrode system inside the chamber. The bakes were performed with a turbo pump attached to the inline valve on the big cube. Note that much of the baking of the MOT chamber and the science side were done separately. This was possible because of the differential pumping tube and the gate valve in between them. The baking procedure for the MOT chamber is much less involved as that chamber is rather simple. The only complexity of the MOT chamber bake was the activation of the dispensers for K and Rb during the bake.

In total, the science chamber was baked for more than 3 weeks, though the temperature was changed a few times during this span. The pressure was measured using the ion pump and a nude ion gauge near the ion pump. We baked near 200 °C for almost a week, and then near 300 °C for more than 10 days. We then went down to 200 °C for a few days, and then down to just over 100 °C for a few more days. The final bake at $\approx$100 °C was used for the activation of the NEG pump and the Ti-sub pump.

The activation procedure for the NEG pump and the Ti-sub was rather complicated, since the goal was to progressively clean the dirt from one and the other in an alternating fashion. The procedure was as follows:

- Turn off the ion pump (the ion pump was mostly off during the previous baking, but occasionally it would be on, especially at the lower temperature during the final bake)
- Activate the NEG pump for 45 min total as initial cleaning (not continuously on because gas load is too high, but continuously up to 33 min)
- Wait a few days for the baking and pumping to remove the extra gas load from the NEG pump
- Activate the NEG pump again for a total of 40 min
- While the NEG pump is on, clean the Ti-sub pump filaments by running 25 A through each of them in turn for 120 s each
- Then, while the NEG is still on but after the Ti-sub pumps turn off, flash the ion pump on five times to clean any dirt off its electrodes
- Then, right before turning off the NEG pump, activate each of the Ti-sub filaments in turn to 33 A for 10 s each
- Finally, turn off the NEG (everything is off now, except the turbo pump), and wait for the pressure to fall below 10^{-8} Torr as measured on the turbo pump
- Decrease the bake temperature from 130 to 100 °C, as a first step in returning to room temperature
- Next activate the Ti-sub pump filaments to 48 A for 45 s each
- Wait one day for the pressure to stabilize, still holding at 100 °C
- Flash the ion pump on five times to clean any dirt off its electrodes
- Cool the chamber down to room temperature with everything off except the turbo pump
- Turn on the ion pump permanently when the chamber has reached nearly room temperature

The pressure reading on the ion pump after this entire procedure and after closing the inline valve to the turbo pump was typically 1.3×10^{-11} Torr (1000 mbar = 760 Torr), which is higher than we wanted.

8.6.4   High-Potting the Ion Pump

We also found that the current (pressure) reading on the ion pump was unstable, and frequently had spikes to several 10^{-11} Torr that were random and unphysically fast. Moreover, we could cross reference the pressure reading with the nude ion gauge, which did not show any of these spikes. To understand this, it is important to consider the current measured by the ion pump. The current is converted to pressure based on the assumption that its only contribution is from the ionization of atoms and molecules that move into the pump. This is called the discharge current.

However, current could also be drawn by the so-called secondary electron currents where electrons can be ejected from the cathode plates. This current increases when the ion energy increases, i.e., the operating voltage of the pump electrodes increases. Another current source is leakage currents, where current can flow between the pump anode and cathode. This can happen when coatings develop on the ceramic elements in the pump, and thus behave like a resistance between high voltage and ground. Yet, another possibility is field emission currents, which are caused by large electric field gradients. The most common cause of such currents is spiky burrs developing on the electrodes, which emit electrons in a similar fashion to the conditioning discussion above.

Such irregular current spikes on the ion pump suggest problems from these latter two mechanisms. These problems are common in ion pumps as they age, but we believe that it happened to us during the pump activation procedure described above. Even though the ion pump was off for most of the sequence, some of the initial activation of the NEG pump was done with the ion pump on, during which it observed gas loads as high as $\approx 10^{-5}$ Torr. The lifetimes of ion pumps are typically given in pressure-hours units, and operating at such high pressures even for short times really cuts into the lifetime of good operation of the pump. Note that the pumping speed of the ion pump may not be affected at all by these issues, but we want to be able to use the pump as an indication of the pressure. The nude ion gauge requires heating to activate, and thus it artificially inflates the pressure and cannot be used constantly.

In order to remove all the deposits onto the ion pump electrodes, we used a so-called hi-potting procedure, which is built into the Gamma Vacuum pump controller for Gamma Vacuum ion pumps. The principle of hi-potting is identical to conditioning, where the current at a given voltage can be minimized by conditioning the electrodes at higher voltages. The maximum operating voltage of the ion pump is 7.5 kV, and so hi-potting is done by increasing the voltage to 10.5 kV for tens of seconds before going back down. We used this procedure a few times over several months. Moreover, we operate the ion pump at 5.5 kV to limit the leakage and field emission currents. By using these tools, the current drawn by the ion pump decreased enormously, and the pressure it calculates is limited by its current resolution to be $P \ll 10^{-11}$ Torr.

8.7   Design of the ITO Coatings

I now discuss the ITO coatings and their designs. We have always used ITO-coated plates that went around the cell in the old apparatus experiment. The coatings were ≈ 40 nm thick, and there was substantial absorption and reflection as a result. However, the old cell was uncoated and also had $\approx 5\%$ reflection.

8.7.1   Coating Thickness Requirements

There are two factors that determine the thickness of ITO that we need on the plates in the science cell. The first is the resistance of the coating, which we model as being in series with the capacitance between the plates. We want to be able to ramp on the electric field in ~100's of ms such that it is adiabatic with respect to the optical trapping frequencies. Therefore, we want the RC time constant to be at a similar level. Since $\tau = RC$ and the capacitance between the plates is $C \approx 0.1$ pF, the resistance of the coating for which $\tau = 100$ ms is $R = 10^{12}$ Ω. In this regard, the resistance of the transmission line is not a concern, and so we typically put 100 MΩ resistors in series with the plates simply to reduce the current, as discussed above.

The other requirement is that the coating be at least ≈100 atomic layers thick such that it is robust to microscopic sources of damage such as alkali atoms or dust which could be burned by an incident laser beam. An atomic layer is roughly 0.1 nm thick for ITO [9], and so a coating thickness of 10 nm is sufficiently thick. The coatings that we used in the old apparatus were 40–50 nm thick, but since the absorption increases with thickness we would prefer to use a thinner coating. Moreover, it becomes more difficult to reduce the reflection by pairing ITO with an AR coating when the ITO coating is too thick. Therefore, we chose to use a coating thickness of 10 nm, and the resistance across the plate that results is ≈800 Ω.

8.7.2 Interaction of ITO Coatings with Alkali Flux

Another important issue with ITO coatings is their known reactivity with alkalis [1]. We could have tried to use an Ni thin film coating instead, but the reflection and absorption would be much higher than ITO. To estimate the reactivity and lifetime, we expect with an ITO coating given the atomic flux in the science cell ($\approx 10^5$ atoms/second), we can measure the lifetime of a test coating in high K vapor pressure environments. Before and after images of such a test plate are shown in Fig. 8.27. These measurements were done by preparing a vacuum chamber that has a K source in it which is not isotopically enriched. We measure the pressure in the chamber as a function of the current through the K dispenser, which controls its temperature. We can also verify the presence of K spectroscopically by performing an absorption measurement when the pressure is high enough. We can also heat

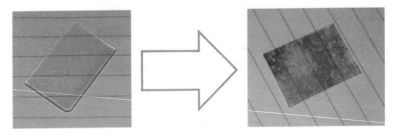

Fig. 8.27 Images of an ITO-coated plate before and after exposure to a high K flux

the vacuum system to help control the pressure of K, which has a very low vapor pressure at room temperature and thus deposits on the walls.

By measuring the pressure P and the temperature T of the vacuum chamber, we can calculate the flux $\dot{N}$ of atoms onto the ITO-coated plate, which has an area A. This can be derived from the ideal gas law, for which we get

$$\dot{N} = \frac{P \cdot A}{k_B T} \sqrt{3k_B T / m_K}, \tag{8.1}$$

where m_K is the mass of K and k_B is Boltzmann's constant. The latter factor is simply the velocity v through one of the axes of the plate that comes from turning the volume in the ideal gas law into $A \cdot x$ and taking the derivative of both sides, giving $\dot{x} = v$ and $\dot{N}$. We measure the lifetime of the coating in two cases: $P = 10^{-7}$ Torr and $T = 150\,°C$, and $P = 10^{-8}$ Torr and $T = 23\,°C$. The lifetime in these two cases is $\tau = 10(2)$ and $>200\,h$, respectively. We expect that the reaction has a strong temperature dependence, so operating at room temperature helps our case significantly.

We can now extend this analysis to the flux we expect from cold atoms being dropped onto the plate roughly every 30 s. The above calculation for the conditions of these tests gives a flux of $\dot{N} \sim 10^{12}$ atoms/second over the area of the plate. Given that we will drop $\sim 10^6$ atoms every 30 s, the flux through whatever area of the plate the atoms fall on to (they will expand over $\approx 5\,mm^2$ after a resonant probe pulse or while very hot at the beginning of optical evaporation) is $\dot{N} = 5 \times 10^4$ atoms/second. This flux is a factor 10^7–10^8 smaller than the measurements, and so we expect the lifetime to be larger than the measured lifetime by a similar factor. Therefore, we estimate that the lifetime of our ITO plates will be $\tau \sim 10^8$–10^{10} hours if we ran the experiment 24/7. This is a very long time, and so we are confident that ITO should work in our system. In other experiments where a Zeeman slower constantly shoots a beam of hot alkali atoms into the chamber containing ITO, the issue of atomic flux onto the ITO coatings becomes a bit more serious.

8.7.3 Pairing the ITO Coating with an Anti-Reflection Coating

Since 2011, all the research we have done has been in at least a vertical lattice, and much of it has been in a full 3D lattice. We expect this trend to continue, and several experiments such as layer selection and evaporative cooling rely heavily on a lattice in the vertical direction. Since the vertical beam will go through many surfaces at small (or zero) angles, we must be mindful of the standing wave contributions from reflections off of all the surfaces. Since there is no well-defined phase relationship between all of these reflections and the main reflection from the dichroic high reflection (HR) plate after the cell, the reflections cause an unstable superlattice which causes enormous heating and destroys BECs very quickly. See Steven Moses'

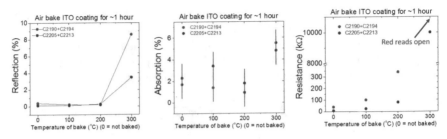

Fig. 8.28 Baking a paired ITO–AR coating in air to various temperatures for 1 h. The plots from left to right show the reflection, absorption, and resistance, respectively

thesis for a more thorough discussion of this effect and a calculation of how the superlattice strength scales with the angle of the beam.

Therefore, we want to pair the ITO coating with an AR coating to minimize the reflection of the ITO-coated plate, particularly at $\lambda = 1064$ nm. Since the ITO layer must be the outer layer, we will put an AR coating on the substrate and then put the ITO coating onto the AR coating. The ITO coatings and AR coatings were developed together by the same company, VacuLayer Corporation in Ontario, Canada. The AR coating was designed specifically for the ≈ 10 nm thickness of ITO as discussed above.

The first AR coating to be paired with ITO was made of MgF_2, and we found that it underwent chemical reactions with ITO when baked in air or vacuum. Figure 8.28 shows that the optical properties through the plate changed significantly when the baking temperature of coated substrates approached 300 °C, even when only baking for 1 h. Moreover, the resistance of the coating increased by 1000× at 300 °C. These observations suggest that the ITO coating is getting destroyed during the bake; however, it could be either that the ITO coating is unstable due to how thin it is or that the ITO-MgF_2 interface is unstable.

To address this question, we performed a test where we baked two substrates with just an ITO coating, where one was 10 nm thick and the other was 150 nm thick. We found that the ITO coating was stable for both thicknesses when baked in vacuum, even up to 300 °C for many days. Therefore, the problem must be coming from the ITO–MgF_2 interface, and we suspect that the following chemical reaction between the oxygen in ITO and the AR coating is being activated at ≈ 200 °C: $O + MgF_2 \rightarrow MgO + F_2$. Upon researching this issue, we found that similar problems have been observed where atomic oxygen in the atmosphere degrades MgF_2 AR coatings on spacecrafts via this reaction pathway [6].

Therefore, we clearly need to use a different AR coating, and we next turned to SiO_x coatings, where $x = 1$–2. VacuLayer prepared such an ITO–AR pairing for us where the ITO coating is ≈ 10 nm thick, and the measured optical properties before any baking are shown in Fig. 8.29. Using this combination, they were able to minimize the reflection at $\lambda = 1064$ nm to $\approx 0.2\%$, and the corresponding transmission was 98%. Both of these numbers are excellent, and they suggest that

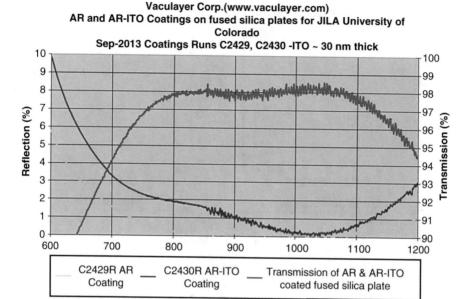

Fig. 8.29 An ITO/AR coating pair using SiO and an ITO thickness of 10 nm. The two different colors of the reflection data are two different coating runs

the absorption is <2%. Note that the other side of the plate has an AR coating which is different than the coating paired with ITO.

Now, we need to bake these substrates as before to see if the coatings survive. We baked the ITO/AR-coated plates to 300 °C in vacuum for up to 6 days, and the results are shown in Fig. 8.30. We can see that while the optical and electrical properties of the coatings change slightly, there are no dramatic changes indicating damage to either coating. In fact, the resistance of the ITO coating actually decreases by ≈20% within a few days and then appears to saturate. This suggests that the ITO coating changes irreversibly as the bake progresses. We suspect that the chemical composition is changing, and our first guess would be that the indium is evaporating slightly during the bake. Since ITO is a semiconductor material, even minute changes to its chemical composition can change its band gap, which would affect the index of reflection at $\lambda = 1064$ nm and its electrical resistance.

It is important to note that the changes in these quantities appear to saturate within 6 days or so. This is corroborated by the observation that after all the final baking procedures discussed above, these properties are still the same as the values after the 6-day test. The reflection has increased from ≈0.5% to ≈2%, and the absorption has increased from ≈2% to ≈5%. Both of these changes will have deleterious effects on the vertical lattice path, and it would be highly advantageous to try to avoid them. While the change to Re(n) could be accounted for using the AR coating if it was understood well enough, there is nothing we can do

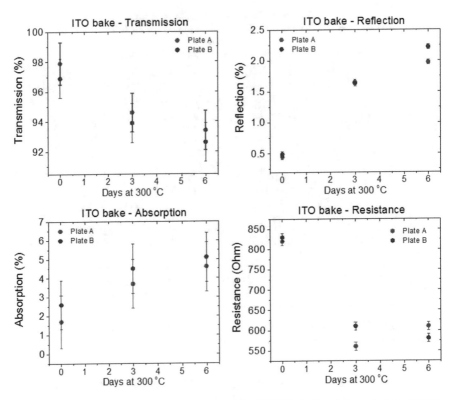

Fig. 8.30 The optical and electronic properties of the ITO/AR coating during a bake to 300 °C. The four plots show (clockwise from top left) the transmission through the substrate, reflection off the ITO/AR coating, absorption of the substrate and coatings, and the electrical resistance across the length of the ITO coating

about the changes in $\text{Im}(n)$. Fortunately, the tests of high intensity on the coated substrates discussed above were also performed for plates which were baked to high temperatures for extended periods. This shows that the increased absorption and reflection still allows us to operate the vertical lattice as desired, but we need to be even more aware of super-lattice effects since the reflections are stronger.

8.8   Optical Qualities of ITO Plates and Cell Windows

The substrates for the ITO-coated plates and the cell windows were produced by Lattice Electro Optics in California. The glass cell was assembled by Precision Glassblowing Inc. here in Colorado. Their procedure is based on machining the quartz skeleton which holds all the substrates, and then frit-contacting the substrates and the quartz tube while pulling vacuum at high temperature. This type of contact uses glass powder (called frit) between the quartz frame and the substrate as a glue when the powder melts at ≈1000 °C. The other viable contacting technique is called optical contacting, and requires that the contacting surfaces are polished to better than λ/10, at which point Van der Waals forces cause the two surfaces to fuse together. This type of fusing can be done at only a few hundred °C, but is much more expensive since super-polished substrates are required.

8.8.1 Anti-Reflection Coatings on the Cell Windows

The frit-based fusing technique requires that the entire cell be heated to ≈1000 °C, and so the AR coatings must be designed to survive such temperatures. Most thin film companies cannot make such coatings, and only two companies in the USA have worked with Precision Glassblowing. One such company is Lattice Electro Optics, the company that provided the substrates for the cell windows and ITO-coated plates. The initial plan for our cell was that they would do the coating, but they backed out in the middle of the job, and so Precision Glassblowing had to turn to the other thin coating provider.

Figure 8.31 shows a photospectrometer scan of the first version of the AR coating before and after baking to the cell-assembly temperature. The blue curve is before baking, and the orange is after baking. Clearly, the coating irreversibly shifts blue by ≈150 nm as a result of the bake. The coating is broadband from ≈750–1100 nm, but it is designed to be optimized at $\lambda = 1064$ nm. To correct for the blue-shift which results from the baking, the coating manufacturer had to shift the coating red

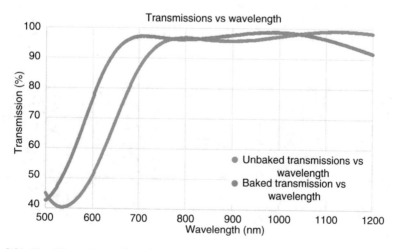

Fig. 8.31 The AR coating on the cell windows before baking to the cell-assembly temperature. The blue curve is before baking, and the orange is after baking

by this same amount. After doing so, the coating beautifully matched the desired wavelength range. Note that the transmission at $\lambda = 1064$ nm is $>99\%$.

8.8.2 Optical Flatness

For high-resolution imaging and addressing as well as for low-wavefront-error optical lattice beams, it is important that the optical substrates of the cell windows and the ITO-coated plates have excellent surface flatness and figure. The flatness refers to bowing, twisting, or wedging due to differential stress or an inhomogeneity in the polishing procedure. For example, the differential pressure between the two sides of the window will cause a r^4 flexure as discussed in Chap. 7. The surface figure refers to how microscopically smooth the surface is, and this of course depends enormously on the polishing procedure.

These quantities can be measured with a Fizeau interferometer which analyzes the interference between a pair of beams, where one of the arms goes through the substrate of interest. The resulting Newton rings can be analyzed to infer the detail of the substrate. The results are shown in Fig. 8.32, where a qualitative difference in the type of aberrations is apparent. The left image (window 1) shows a clear wedge across one axis of the window, which causes coma aberrations. The right image (window 2) shows a roughly cylindrically symmetric Newton ring resulting from bowing, which causes distortion (i.e., variation of the magnification of the imaging system across the image plane) and spherical aberration.

The traces in these two figures are cuts through the interference pattern across the substrate, and they can be used to quantify the surface quality. The larger feature across the substrate is a measure of the surface flatness, which is $\lambda/5$ for both windows at $\lambda = 780$ nm. The RMS amplitude of the smaller oscillations of this envelop is a measure of the surface figure, which is $\approx \lambda/10$ for both. The

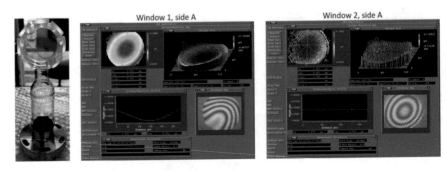

Fig. 8.32 Fizeau interferometry of the cell windows

quality of the substrates for the ITO/AR coatings was measured to be $\approx \lambda/5$. These measurements were not repeated with the chamber under vacuum or after baking, but we expect the surface flatness to be $\lambda/5 - \lambda/4$ for all the substrates after all baking. Such values are typical even for quantum gas microscope experiments where the imaging system has NA= 0.65 [10], and an optical system with an optical path difference (OPD) of $\lambda/4$ is considered a perfect system from a geometrical optics point of view, and this criteria is called the Rayleigh limit [3].

8.9 Testing the Objective

We tested the resolution of the objective with two techniques. The first test used an Air Force test target, which is a glass substrate with sets of three lines progressively smaller in a spiral pattern. A portion of the test target is shown in Fig. 8.33, and the

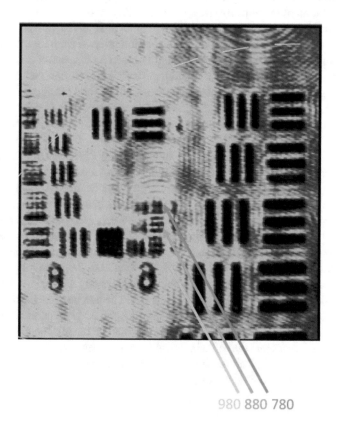

980 880 780

Fig. 8.33 Testing the microscope objective by imaging an Air Force test target onto a profile camera. The smallest line spacing that can be resolved is 880 nm, suggesting that the resolution is $\approx$850 nm

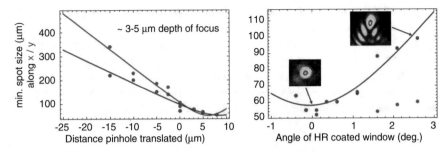

Fig. 8.34 Characterizing the resolution limit of the objective using a small pinhole. The left plot measures the depth of focus by translating the pinhole relative to the object plane. The right image measures the effect of tilt of the 2 mm thick plate between the pinhole and the objective

smallest line sets that can be resolved are in the middle of the image. The target is imaged onto a profile camera using a magnification large enough that sizes limited by the resolution of the objective (r_{min}) are magnified to occupy many pixels on the camera. The line set of spacing 770 nm cannot quite be resolved, but the 880 nm line set can be resolved easily. Therefore, the resolution inferred from this measurement is ≈850 nm, which is similar to the expectation based on NA = 0.53.

A more rigorous and useful measurement of the resolution can be done by illuminating a pinhole much smaller than the resolution, and placing it in the object plane. In this limit, the image of the pinhole is equivalent to the image of a single atom (point-source scatter), and this approach allows us to measure the point spread function of the objective. We used a pinhole of diameter $D = 500$ nm, and some characterizations of the resolution are shown in Fig. 8.34. The measurement is performed in the same way as the test target, where a second lens is used to image the pinhole onto a camera with sufficiently high magnification to avoid pixelation. Note that both of these tests are done with glass pieces between the object plane and the objective which match the thickness of the elements used in the new apparatus.

The plot on the left shows the depth of focus, which is measured by translating the pinhole along the imaging axis. The size in the image plane gets larger as the object is moved out of focus, and the fit is to a Gaussian Rayleigh range function. The depth of focus is defined when the object becomes larger by $\sqrt{2}$, which is ≈3–5 μm, consistent with what we expect for this NA. The plot on the right is looking at the effect from tilting along the x-axis a 2-mm-thick glass plate between the object plane and the objective. This test is designed to simulate the relative angle between the ITO-coated plates and the cell window. A large tilt angle causes significant coma aberrations, as shown in the inset images on the camera. This data suggests that the measured tilt angle of 0.2° should be negligible.

8.10   Magnetic Fields

Now, I will discuss all the coils that go around the science cell. Measurements of magnetic fields were performed with a Gauss meter that was calibrated to zero field using a μ-metal shield.

8.10.1   Testing the Bias Coils

Figure 8.35 shows measurements of the voltage and magnetic field as a function of the current in the coils. The left plot measures the resistance of the coils, and the right plot shows the magnetic field measured with a Gauss meter as a function of the current. The measured data points and fit are shown in black, and the calculated points and theoretical fit described in Chap. 7 are shown in red. The two are in excellent agreement. The K-Rb Feshbach resonance is at 547 G, which can be reached with $\approx$220 A. At 1005 G is a narrow Feshbach resonance for Rb, which could be useful to measure double occupancy of Rb by making Rb_2 Feshbach molecules. This field can be reached with $\approx$400 A. Another important field is $\approx$1260 G, at which the excited rotational states $|1, 1\rangle$, $|1, 0\rangle$, and $|1, -1\rangle$ are degenerate [8]. This field can be reached at $\approx$480 G, which is still within reach of our high-current power supply.

We also measured the heating as a result of going to such high currents. Figure 8.36 shows the steady-state operating temperature measured in a few places as a function of the current. The coils are hollow core and have water flowing through them, so each of the 16 turns should be at roughly the same temperature.

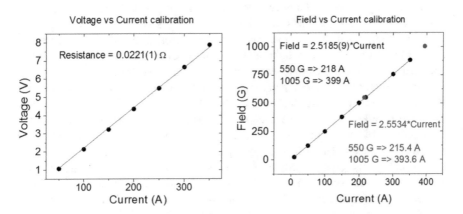

Fig. 8.35 Measuring the resistance and calibrating the magnetic field from the bias coils. The left plot measures the resistance of the coils, and the right plot shows the magnetic field measured with a Gauss meter as a function of the current. The measured data points and fit are shown in black, and the calculated points and theoretical fit are shown in red. The two are in excellent agreement

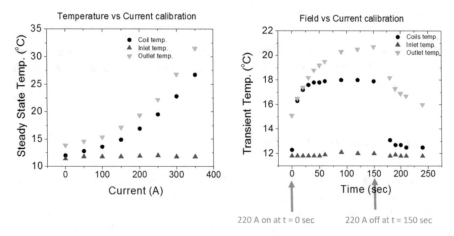

Fig. 8.36 Thermal effects on the coils as a function of the applied current. The left plot shows the steady-state temperatures measured on the water inlet to the coil (red), on the top turns of the 4 × 4 turn coil (black), and on the water outlet (green). The right plot shows the transients measured in the same places when 220 A is stepped on, and then later stepped off

The temperature of the coils enters a linear regime at above ≈250 A, where the steady-state temperature increases by ≈5 °C for every 50 A. This trend suggests that at even 500 A the coil temperature would be <40 °C, which is acceptable for hold times of several seconds.

We can also look at the transients to see how long it takes to reach the steady-state temperatures. This is shown in the right plot of Fig. 8.36 for the same three temperature probe locations. This shows that for 220 A (550 G, where the K-Rb resonance is), the heating and cooling occurs on a $\tau \approx 10\,\text{s}$ timescale. This timescale will be roughly the same at any current, only the amplitude will vary. Typically, we only need to apply large fields for a few seconds, and so this heating in negligible even to 500 A. Such thermal cycling of several seconds per 30–40 s is certainly acceptable, but mechanical fatigue will eventually set in, and the glue between the coils will have to accommodate slow mechanical drift over several years.

It is important to consider the homogeneity of the field to identify how much field curvature exists at the center, and to estimate how sensitive we will be to exactly how the coils are oriented around the science cell. Figure 8.37 shows the curvature of the z-component of the field in the axial and radial directions of the coils as measured with a Gauss meter. The field is fit to a quartic polynomial, where the only fitting parameters are the quartic term amplitude and the offset. These estimates suggest that the field variation over the size of the cloud (100 μm) in the center is ≈6 mG in both directions at 250 G, which is roughly 2 parts in 10^5. The figure of merit is the spread in the K transition $|9/2, -9/2\rangle$ to $|9/2, -7/2\rangle$ at 550 G, where the field variation over the cloud would be ≈10 mG. The sensitivity of this transition is ≈40 kHz/G, and thus this field variation will broaden the transition to 400 Hz. This

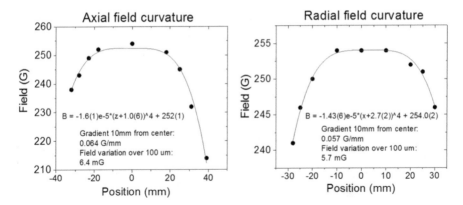

Fig. 8.37 Curvature of the z-component of the bias field in the axial and radial directions of the coils. Measurements were made with a Gauss meter oriented along the z-direction mounted on a 3D translation stage

is a smaller effect than the field noise due to the servo quality and is also smaller than the Rabi frequency of the K 80 MHz rf coil, which is ≈ 10 kHz.

8.10.2   Testing the Gradient/Quantization Coils

We can perform similar tests of the small, magnet wire coils inside and concentric with the bias coils. These coils will be used for quantization coils, fast magnetic field sweeping coils, and also for gradient coils. The direction of the current in one coil can be switched relative to the other coil using an H-bridge, which allows us to change them from a Helmholtz configuration to an anti-Helmholtz configuration. The field from the coils in the Helmholtz orientation and the gradient from the coils in the anti-Helmholtz configuration as a function of current are shown in Fig. 8.38.

The field in Helmholtz configuration is in excellent agreement with the predicted value, giving 6 G/A. We can apply up to ≈ 10 A for a few seconds, allowing us to apply fields of around 60 G. These fields can be switched very quickly, allowing for fast magnetic field sweeps across the Feshbach resonance. We configure the current directly such that this field adds with the field from the bias coils. Then, we can initialize the system above the resonance, where both the bias coils and the fast Helmholtz coils are on. We can set the bias coils to the desired field below the resonance, and then when we quickly turn the current in the small Helmholtz coils to zero, we sweep across the Feshbach resonance. The high-field bias coils have a more stable servo and their field profile is more uniform. These factors are only important below the resonance when we perform rf manipulation of K, and in this way the small Helmholtz coils are off for such manipulation.

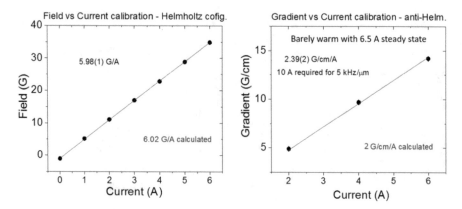

Fig. 8.38 Calibrating the field and field gradient from quantization/gradient coils. The left plot shows the field as a function of current with the coils in the Helmholtz configuration. The right plot shows the field gradient as a function of current with the coils in anti-Helmholtz configuration

The calibration for the coils in the anti-Helmholtz configuration gives 2.4 G/cm/A, and thus we can reach 25 G/cm with these coils, which is sufficient to levitate atoms against gravity. We use these coils for spin-dependent detection to measure the relative spin populations in the clouds of atoms. We can also use this quadrupole field to compress the clouds of atoms in the science cell that are spread out along a single laser beam. We optically transfer atoms into the science cell in the focus of one beam, and this gradient from the quadrupole field allows us to compress the cloud in order to enhance the subsequent loading into a crossed dipole trap. This will be discussed in Chap. 9.

8.11 Electric Field Voltage Servos

Any high-fidelity manipulation of rotational states in a large electric field will require excellent electric field stability. If the Rabi frequency of the rotational transitions is $\Omega \approx 10$ kHz, we need the field fluctuation during the pulse to result in frequency shifts <1 kHz. As discussed in Chap. 7, the shifts of the $|0, 0\rangle$ to $|1, 0\rangle$ transition are $\approx 100–200$ kHz/(V/cm), and thus we need the field to be stable to $\Delta E \approx 10$ mV/cm. At an electric field of $E \approx 10$ kV/cm, the required stability is $\Delta E / E = 10^{-6}$. Accordingly, we want to be able to stabilize the voltages applied to the electrodes at the part-per-million (ppm) level.

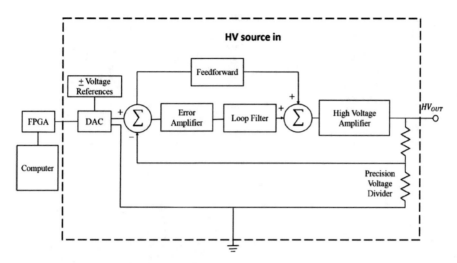

Fig. 8.39 A schematic of the servo circuit to stabilize the voltages after the high voltage amplifiers. Reproduced from Reference [5]

8.11.1 Stabilizing Large Voltages

The design, construction, and testing of the servos to stabilize the voltages from the high voltage amplifiers was done by Terry Brown of the JILA instrument shop and Jamie Shaw, an undergraduate writing his honors thesis in our group. More information on these servos can be found in the thesis of Jamie Shaw [5]. A schematic is shown in Fig. 8.39. The input voltage to be amplified is from a digital-to-analog converter (DAC), and it is derived from a field programmable gate array (FPGA) connected to a computer. Moreover, the DAC uses a temperature-stabilized voltage reference such as an LM399. The requirement for the stability of this DAC voltage is as stringent as the stability of the high voltage.

This analog voltage needs to be compared with the voltage after amplification, and we do this by using an ultrastable voltage divider. The voltage divider was designed carefully to minimize thermo-couples (connections between different metals), but a larger concern is the stability of the resistors. We used ultrastable Vishay resistors (Z-foil type) with a temperature coefficient of 0.05 ppm/$^\circ$C. Since we want to reach ppm stability, we must be careful of the temperature stability of the resistors. Thermal effects are generally slow, but since the voltage divider actually dissipates current we must be aware of the heat load on the resistors during measurement. While this is an important factor, we believe that this is not a major limitation at the few ppm level.

The divided voltage is summed in with the input voltage, and the sum goes into an error amplifier and the loop filter, and is then summed with a feed forward voltage. The output of that goes to the high voltage amplifiers. The grounding of all of these

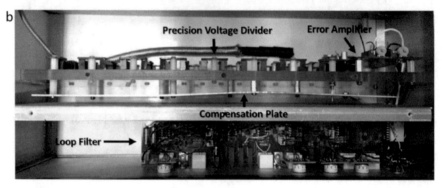

Fig. 8.40 The circuit for stabilizing the voltages after the high voltage amplifiers. (**a**) The voltage divider to compare the voltage after the amplifiers to the input voltage. (**b**) The Servo box, with the voltage divider on the top and the loop filter servo on the bottom. Note the compensation plate below the voltage divider. This was painstakingly adjusted to optimize the servo performance and bandwidth. Reproduced from Reference [5]

elements is critical, as shown in Fig. 8.39. The voltage dividers and servo circuits are shown in Fig. 8.40. Below the voltage divider is a compensator plate which is used to improve the servo bandwidth. It does this by coupling signals back into the error amplifier board connected to the voltage divider that would have been lost due to capacitance between the resistors and the nearby metal box [5].

8.11.2   Realizing ppm Voltage Stability

In order to characterize the quality of the servo, we can begin by looking at the stability of the DAC voltage reference, which is shown in Fig. 8.41a. These measurements were performed with a 6-digit digital multi-meter (DMM), and the measurement was broken into two ranges to get the best data from the DMM: 1 V and 10 V. The black and red data shows the stability of the DAC voltage as a function of the voltage. Their fractional stability improves enormously as the voltage

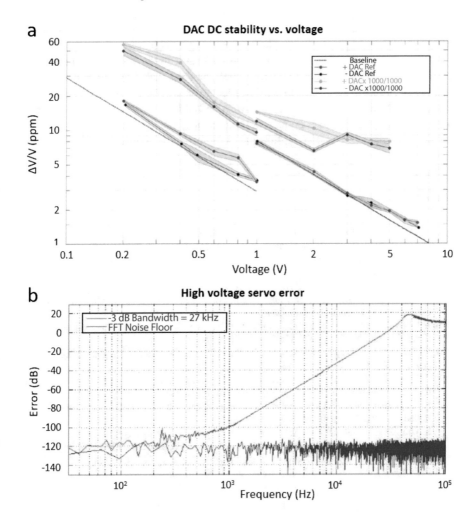

Fig. 8.41 Measuring the voltage stability of the LM399 reference and the servo output. (**a**) The voltage stability versus the voltage. This measurement was broken into two ranges to get the best data from the DMM: 1 V and 10 V. (**b**) The error signal from the servo loop as a function of frequency on an FFT machine. This curve shows that the servo bandwidth is several kHz. Reproduced from Reference [5]

approaches the LM399 reference voltage of $\approx$7 V. Both of which are reasonably close to the noise floor, which is shown as the black line.

We performed an open loop measurement of the stability of the high voltage amplifiers. This was done by amplifying the DAC voltages (1000$\times$), and then dividing it by 1000 with the voltage divider. The blue and green data show this divided voltage and suggest that the high voltage amplifier is $\approx$20 ppm worse at most voltages. The amplified voltage follows the DAC voltages, suggesting that its noise is not strongly dependent on the voltage.

Now, we can turn on the servo loop and see how effectively it removes the noise from the high voltage amplifier. The loop filter error signal is analyzed on a fast Fourier transform (FFT) as a function of the frequency in Fig. 8.41b. The noise floor of the FFT is shown in blue, and the red shows the error. This suggests that the servo bandwidth is 10's of kHz. Therefore, we expect to have several-ppm-level stability at $\pm 7\,$kV potentials up to frequencies of around $\approx 10\,$kHz. Such values should enable high-fidelity rotational state control in large electric fields, which can be used for quantum magnetism and non-equilibrium many-body dynamics.

References

1. R. Daschner, R. Ritter, H. Kübler, N. Frühauf, E. Kurz, R. Löw, T. Pfau, Fabrication and characterization of an electrically contacted vapor cell. Opt. Lett. **37**(12), 2271–2273 (2012)
2. N. Fitch, Traveling-wave stark-decelerated molecular beams for cold-collision experiments, PhD thesis, University of Colorado, Boulder, 2013
3. R. Guenther, *Modern Optics* (Wiley, New York, 1990)
4. E. R. Hudson. Experiments on cold molecules produced via stark deceleration. *PhD thesis, University of Colorado, Boulder*, 2006.
5. J.C. Shaw, External electric fields: a new tool for controlling ultracold polar molecules, BS thesis, University of Colorado, Boulder, 2015
6. D. Speckman, D. Marvin, J. Matossian, N. Ianno, W. Stuckey, Atomic oxygen testing of mgf$_2$ coatings, in *IEEE Conference Proceedings*
7. B.K. Stuhl, Ultracold molecules for the masses: evaporative cooling and magneto-optical trapping, PhD thesis, University of Colorado, Boulder, 2012
8. S.V. Syzranov, M.L. Wall, V. Gurarie, A.M. Rey, Spin-orbital dynamics in a system of polar molecules. Nat. Commun. **5**, 5391 (2014)
9. S. Umadevi, V. Ganesh, S. Berchmans, Liquid crystal (LC) monolayer on indium tin oxide (ITO): structural and electrochemical characterization. RSC Adv. **4**, 16409–16417 (2014)
10. C. Weitenberg, Single-atom resolved imaging and manipulation in an atomic Mott insulator, PhD thesis, Ludwig-Maximilians-Universität München, 2011

Chapter 9
Experimental Procedure: Making Molecules in the New Apparatus

Most of the laser systems for the experiment are the same between the old apparatus and the new apparatus, so to change to the new apparatus, we moved the old vacuum chamber off of the main optical table for the experiment, and on to a smaller optical table for testing and building. The new chamber was then moved from this table to the position on the main experiment table where the old chamber was previously. The moving of the chambers was done with a mobile jacking lift platform which was outfitted with a large aluminum breadboard. The two sides of the flexible bellows were braced on both chambers when moving them. Figure 9.1 shows the entire chamber-moving process.

In this chapter, I discuss the experimental procedure for creating gases of ultracold atoms and molecules in this new apparatus. To integrate the new apparatus into the experiment, we first had to remove the old apparatus as shown in Fig. 9.1. A full view of the new apparatus is shown in Fig. 9.2, and the different regions for atom manipulation and trapping are indicated.

9.1 Dual Species MOT with Gray Molasses Cooling of ^{40}K and ^{87}Rb

The first section of the chamber is the vapor MOT cell, where we load a bichromatic MOT for both K and Rb from a background vapor. The K dispensers are enriched to 14% abundance of ^{40}K, above the natural abundance of 0.012%. The pressure in this regime is maintained to give a $1/e$ MOT fill time and magnetic trap lifetime of roughly 2.5 s. The MOT beams have a $1/e^2$ radius of 1.5 cm and power of $\sim$200 mW, and typical MOT sizes are $N_{Rb} = 2 \times 10^9$ and $N_K = 5 \times 10^7$. The quadrupole gradient used during the MOT is $\approx$15 G/cm on the strong axis.

In the new apparatus, we started using laser-induced atomic desorption (LIAD) to reduce the MOT loading time and enhance the MOT size. LIAD is induced by

© Springer Nature Switzerland AG 2018   191
J. P. Covey, *Enhanced Optical and Electric Manipulation of a Quantum Gas of KRb Molecules*, Springer Theses, https://doi.org/10.1007/978-3-319-98107-9_9

Fig. 9.1 Photographs of the process of moving the vacuum chambers. (**a**) Moving the old chamber off of the main experiment table. (**b**) Moving the new chamber onto the main experiment table. (**c**) Installing the new chamber in the position that the old chamber was previously

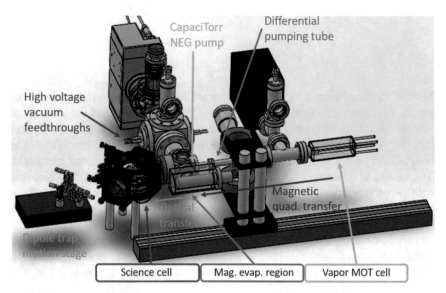

Fig. 9.2 A render of a solid model of the entire new apparatus vacuum chamber. The three chambers used to prepare ultracold atoms and molecules are identified. The large cube where the high voltage feedthroughs couple in large DC voltages is evident, and the high-resolution objective can be seen under the science cell

UV light which facilitates desorption of atoms from glass surfaces [15]. While the efficacy of LIAD depends on many factors such as the atomic species, the type of material in the chamber to which the atoms are adsorbed, and the temperature of the chamber, it works very well in our system. We use a UV diode centered at $\approx$260 nm near the Pyrex MOT chamber as shown in Fig. 9.3. The MOT chamber is heated to $\approx$40–50 °C to prevent K deposits from developing on the glass and obstructing the MOT beams. We turn the LEDs on to 0.5 A for 1 s at the beginning of the MOT loading phase, and in doing so we were able to save $\approx$10 s on the experimental cycle time.

Following the MOT loading, we use a compressed MOT stage (CMOT) [19], and then optical molasses [13] for both K and Rb. For Rb, bright molasses allows us to reach a minimum temperature of 40 μK with small clouds, and 200 μK for maximum density, limited by photon rescatter. For K, bright molasses on the D2

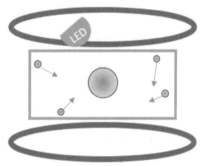

Fig. 9.3 UV LEDs for LIAD of K and Rb in the MOT chamber. The quadrupole coils are shown on the right, and the LEDs are shown in blue. The purple cloud indicates that the LEDs facilitate the desorption of atoms from the walls of the Pyrex chamber

($P_{3/2}$) transition at $\lambda = 766.7$ nm is even less efficient due to the limited hyperfine splitting of the electronic excited state [16], and we reach $\approx 250\,\mu$K. In the new apparatus, we have added gray molasses for both K and Rb to circumvent these issues and cool both to 10's of μK or less.

For gray molasses of K, we use the D1 ($P_{1/2}$) at $\lambda = 770$ nm instead [2, 9], where the hyperfine structure in the electronic excited state is better resolved. As shown in Fig. 9.4, the grey molasses operates as a Raman transition between the two hyperfine ground states with a blue detuning in the electronic excited state. Typical detunings for both species are $\Delta_B \approx 20$ MHz, which are similar to the bright molasses detunings Δ_R. However, we find it best to ramp Δ_B for both K and Rb closer to resonance during the gray molasses time. The scheme for Rb is similar, except the D2 ($P_{3/2}$) can still be used since the hyperfine splitting is well resolved in the electronically excited state.

Since the gray molasses relies on coherent Raman transitions between the hyperfine ground states (see Fig. 9.4), phase coherence between the two lasers is necessary. Therefore, for K and Rb we use EOM phase modulators with sidebands at 1.29 GHz and 6.84 GHz, respectively. Both EOMs are from New Focus, and we drive them with ≈ 35 dBm, which gives ≈ 7–8% of the total power in each sideband. Note that only one sideband is useful for each species, and that it happens to be the opposite sideband for each due to the inversion of the hyperfine structure of ^{40}K. This phase coherence allows us to observe electromagnetical induced transparence (EIT) between the two laser fields as a function of the EOM frequency.

For both K and Rb, we find that the overall molasses number and temperature is optimized when we do a few ms of bright molasses prior to gray molasses. We find that ≈ 8 ms of gray molasses is optimal, during which we ramp the intensity and power of the molasses beams after the first few ms. During these ramps, we decrease the power and the detuning. This approach allows us to reach 60(8) μK for K and 7(2) μK for Rb at the end of the molasses stage. For the Rb gray molasses, there is a ≈ 50% loss of atoms, presumably because the capture velocity is comparable to

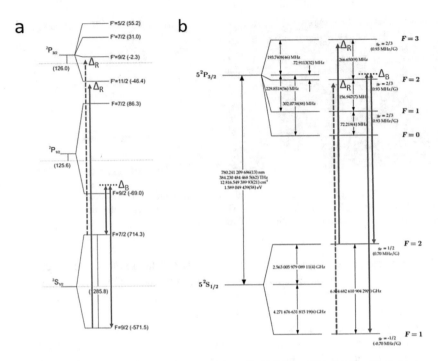

Fig. 9.4 Bright and gray molasses laser detunings for both K and Rb. (**a**) The level structure of K, including fine and hyperfine structure. Note the inverted hyperfine structure for ^{40}K. The bright molasses is red-detuned as shown with the red beams. The dashed beam is the repump. Both have a detuning from their respective electronically excited hyperfine state of Δ_R. The gray molasses is blue-detuned, and operated by a Raman condition with detuning Δ_B, as shown with the blue arrows. For K, the bright molasses uses the D2 state and the gray molasses uses the D1 state. (**b**) The level structure of Rb, including hyperfine structure. Note that only the D2 electronically excited fine structure is shown. The bright molasses is red-detuned as shown with the red beams. The dashed beam is the repump. Both have a detuning from their respective electronically excited hyperfine state of Δ_R. The gray molasses is blue-detuned, and operated by a Raman condition with detuning Δ_B, as shown with the blue arrows. For Rb, the bright and gray molasses both use the D2 state. Note that the level structures in (**a**) and (**b**) are not necessarily to scale

the temperature during the bright molasses that precedes it. Nevertheless, the phase-space-density is significantly higher with the gray molasses for both species.

After molasses, we optically pump Rb to $|F, m_F\rangle = |2, 2\rangle$ (F is the hyperfine quantum number and m_F its projection onto the magnetic field) and K to $|9/2, 9/2\rangle$, which are low magnetic field seeking states, both with a magnetic moment of $1\mu_B$. Then, we load the atoms to a quadrupole trap with gradient on the strong axis of $\approx$40 G/cm. To characterize the efficiency of all these steps, we routinely find that the number loaded into the quadrupole trap after CMOT, molasses, and optical pumping is 80% of the initial number in the MOT. The temperature after loading into the quadrupole trap with gray molasses is $\approx$100 μK for Rb, and $\approx$220 μK for K. The corresponding temperature with D2 bright molasses cooling of K instead is 700 μK.

Note that there is considerable adiabatic and non-adiabatic heating of both species during the loading of the quadrupole trap from molasses.

9.2 Magnetic Transfer and Plugged Quadrupole Evaporation

Once the atoms are loaded into the quadrupole trap, which is composed of anti-Helmholtz coils on a motion track, we move the clouds through the vacuum chamber to the evaporation region using a gradient of ≈ 160 G/cm, as depicted in Fig. 9.2. The coils were aligned to the chamber axis using axial imaging through the differential pumping tube, as shown in Fig. 9.5. We perform forced microwave evaporation in the quadrupole trap driving Rb from the magnetically trapped $|2, 2\rangle$ state to the anti-trapped $|1, 1\rangle$ state with radiation from a microwave horn at ≈ 6.85 GHz. We sympathetically cool K with Rb [23].

The starting temperature before evaporation is $\approx 200 \mu$K for both species, although this is difficult to measure. We find that the initial number of atoms is improved using gray molasses by $6\times$ for K, and by $2\times$ for Rb. While the number of K atoms in the quadrupole trap before magnetic transfer is similar with or without gray molasses, the lower temperature with it allows for a substantially higher transfer efficiency through the chamber, where a hotter cloud is severely clipped by the aperture of the differential pumping tube in the vacuum chamber. The pressure in the quadrupole region of the vacuum chamber is orders of magnitude lower because of this differential pumping and is expected to $\ll 10^{-11}$ mBar. The lifetime in the quadrupole trap depends on the temperature because of Majorana spin flips at the zero-field center of the trap [1], but was measured to be > 150 s.

While evaporation to 10's of μK with Rb in a quadrupole trap before Majorana loss becomes significant is routine [14, 21], the smaller mass of K and the larger number of magnetic states (larger F) causes significant loss at much higher temperatures. Therefore, we use a blue-detuned optical plug beam that repels the atoms from the magnetic field zero crossing [6, 8, 11]. We use a plug with wavelength $\lambda = 750$ nm, which is ~ 16 nm blue-detuned of the K D2 line, and ~ 30 nm blue of

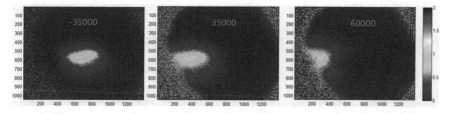

Fig. 9.5 Aligning the cart coils with the chamber axis. The three images here show the cloud in the magnetic coils at three different positions, each with progressively further transfer while at an incorrect angle. This measurement is used to align the track such that the cloud stays centered over the full range

the Rb D2 line. This laser is derived from an M^2 SOLSTIS, which is a titanium-sapphire tunable laser from $\approx$740 nm to $\approx$950 nm capable of delivering $\approx$ 5 W. We use a $1/e^2$ beam radius of 50 μm with a power of $\approx$1 W, giving a repulsive barrier height of 300 μK for K and 200 μK for Rb. We evaporate to a temperature of 16(2) μK in 13 s, yielding $N_{Rb} = 2.0(3) \times 10^7$ and $N_K = 5.0(4) \times 10^6$.

Evaporation of two species using sympathetic cooling often requires careful removal of coolant atoms in the wrong spin state, such as Rb in $|2, 1\rangle$ [22]. In the old apparatus, we used a so-called $|2, 1\rangle$-cleaner in the IP trap as discussed in Chap. 2. The application of such a cleaner using a second rf tone is more complicated in the quadrupole trap where both $|2, 1\rangle$ and $|2, 2\rangle$ become degenerate at the bottom of the trap, although the plug beam ostensibly prevents atoms from spending appreciable time in this region. We spent some time working with a $|2, 1\rangle$-cleaner since we anticipated that it would be important, but in the end we found that the Rb purity in $|2, 2\rangle$ is excellent throughout evaporation, and we never needed the $|2, 1\rangle$-cleaner in the new apparatus. This is because we choose the gradient of the quadrupole trap after optical pumping to be small enough that we cannot efficiently hold $|2, 1\rangle$ atoms, and thus they never appear in appreciable quantity in the quadrupole trap. The K number changes by a factor of $\approx$3–4 throughout the evaporation, but this is expected as a result of three-body loss, collisions with Rb atoms in $|1, 1\rangle$ as they leave the system, off-resonant light-scattering from the plug beam, and other loss mechanisms (a similar factor of K was lost in the old apparatus).

9.3   Optical Transfer into the Science Cell

After evaporation in the magnetic trap, we ramp down the quadrupole trap and the plug beam in 600 ms while ramping up a single optical dipole trap beam with wavelength $\lambda = 1064$ nm. The waist is 46(3) μK, and the final power is 7–8 W, such that the trap depth is $\approx$500 μK for both K and Rb. We align the dipole trap beam (called the transfer beam) onto the plugged quadrupole trap in order to maximize the loading efficiency, but the relative positions of the plug and transfer beam depend on the relative power.

9.3.1   Transfer Beam Alignment onto the Plugged Quadrupole Trap

In the limit that the trap depth of the transfer beam is much larger than the barrier height of the plug (we were briefly using a very large trap depth of the transfer beam), the two together give a net attractive potential. This potential holds all the atoms at the zero-field location (where the plug is aligned), and so there is substantial loss when the transfer beam is at this position. Such a dip is apparent

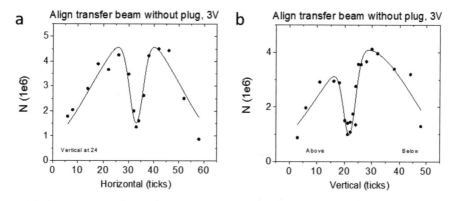

Fig. 9.6 Alignment of the very strong transfer beam onto the plugged quadrupole trap. (**a**) Scanning the horizontal position of the transfer beam with the vertical roughly centered. (**b**) Scanning the vertical position with the horizontal roughly centered. "Above" and "below" indicate the relative position of the transfer beam focus with respect to the quadrupole trap center

for Rb in Fig. 9.6 where the transfer beam position is scanned in the horizontal and vertical direction. The vertical profile is asymmetric because of gravity, and thus we set the transfer beam to be centered in the horizontal and below in the vertical, which corresponds to the best points in Fig. 9.6b.

9.3.2   Lifetime and Heating in the Transfer Beam

Typically, we operate with the transfer beam depth roughly the same as the plug barrier height. In this case, they roughly cancel each other out when perfectly overlapped, and the loss rate of either species at this temperature is small compared to the timescale that both beams are on together. Therefore, we typically do not see these dips when both beams are perfectly overlapped in scanning the transfer beam (as in Fig. 9.6) during typical operation. We find typically 50–60% loading efficiency for both K and Rb. The lifetime in the transfer beam is typically ≈ 10 s, which is limited by pointing noise of the transfer beam. The transverse motion of the beam drives excitations to excited motional states in the trap which causes significant heating which leads to loss. The lifetime was initially ≈ 1 s, and it took some dedicated effort to understand the source of the heating and reduce the pointing noise.

Typical transverse frequencies are $f \approx 1$ kHz, which is a ubiquitous frequency for acoustic noise of small, metal objects such as mirror mounts and posts. Moreover, since the separation between the radial states is not well resolved at these temperatures (1 kHz$\sim$50 nK), atoms can easily be excited multiple times which gives rise to very large heating rates. To characterize these effects, we can drive excitations in the radial degree of freedom using a piezo-actuated mirror in the

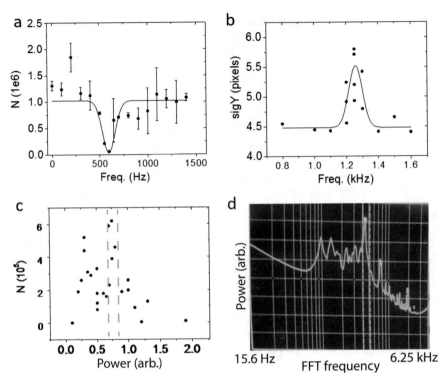

Fig. 9.7 Characterizing the heating of Rb in the transfer beam from pointing instability. (**a**) Pointing modulation using a piezo-mirror, providing a dip at the trap frequency. (**b**) Amplitude modulation creates a dip at roughly twice the trap frequency (depending on the role of anharmonicity). The trap depth is $\approx 150\,\mu K$ for this data, which is lower than we typically use. (**c**) The lifetime in the transfer beam as a function of the beam power as measured by the atom number after holding 2 s. (**d**) An FFT scan of the quadrant photodiode at the focus of the transfer beam. The large peaks match the frequencies of the traps that have large dips in the lifetime

beam path to measure the $v = 0$ to $v = 1$ transition, and amplitude modulation (parametric heating) to drive the $v = 0$ to $v = 2$ transition. Figure 9.7a, b shows the resonances of these two modulations at a slightly lower trap depth, and the width gives an estimate of the anharmonicity sampled by the atoms at such temperatures.

The lifetime of Rb in the transfer beam as a function of power is shown in Fig. 9.7c, and there are obvious resonances at which the lifetime is particularly short. The lifetime at very high trap depth starts to decrease because of three-body loss. We use a quadrant photodiode and an FFT machine to characterize the pointing noise of the transfer beam, and a typical measurement from the initial setup is shown in Fig. 9.7d where the amplitude of the noise is plotted in $mV/\sqrt{Hz}$ as a function of the frequency. In fact, the trap depths in Fig. 9.7c where the lifetime is very short have corresponding transverse trapping frequencies exactly matching the frequencies of the largest spikes in Fig. 9.7d.

To solve this problem, we had to rebuild almost the entire beam setup for the transfer beam. Since the beam has to be too large for 1-in. optics, we are required to use a beam height of two inches above the breadboard and optical transfer stage. We use only stainless steel optics holders and 1 in. pedestals for all optics, so there were not many potential upgrades available with the optics. Most of the optics were on a 1/2 in. thick aluminum breadboard held above the table by long 2 in. diameter pillars. We found this mounting scheme to be particularly resonant at most acoustic frequencies, and since the breadboard is undamped the vibrations would persist for long times. Therefore, most of the optics are now on a honeycomb breadboard that is 4.3 in. thick, under which are two 1-in. thick stainless steel slabs. We use thin sheets of highly damping Viton between the table and the first steel slab, between the two steel slabs, and between the upper steel slab and the breadboard. All of these elements are bolted to the table through the breadboard with long screws, where there are clearance holes through everything.

The motion stage used to be mounted alone on four 2-in. diameter pedestals, but we reduced the pedestal height by 2 in. and we use two 1-in. thick steel slabs here as well, again with Viton between each layer. The stage is bolted to the pedestals through clearance holes in the steel slabs (see Fig. 9.10). As a result of these improvements to the optical surfaces for the transfer beam optics, we now have a lifetime of ≈ 10 s and we do not observe any heating during this time. We are satisfied with this result, and as I will discuss later we have measured lifetimes of ≈ 30 s in this beam in the science cell. Nevertheless, it is still likely the case that the lifetime is significantly less at specific trap depths whose resonant trapping frequencies exactly match some pointing noise, although the predominant peaks that remain are ≈ 100–200 Hz, which is well below any relevant trap depth.

9.3.3 Loss of Atoms During Transfer

With this problem solved, we then go on to translate the focus of this optical trap beam by using an Aerotech magnetic-bearing translation stage (ANT130-110-L-PLUS) holding two mirrors which are between the focusing lens and the vacuum chamber. Since the lens is not on the stage and we do not transfer BECs, we are less sensitive to the smoothness of the transfer process, and thus we do not need to put the stage on a superpolished granite slab. We align the beam incident to the stage and the mirrors on the stage such that the beam translates by less than $10\,\mu$m between the two ends of the travel range (see discussion below on changes to the mirror mounts on the stage). The beam goes into the chamber through the end window of the science cell, and its focus carries the atoms ≈ 17 cm from the quadrupole evaporation region into the science cell (see Fig. 9.2).

Initially, we ran into enormous problems with the optical transfer that took quite a while to understand. We found that in transferring Rb, the number of atoms dropped sharply after transferring 2 cm. We were eventually able to determine that this sharp loss was a magnetic effect since it was significantly less drastic for atoms in $|1, 1\rangle$

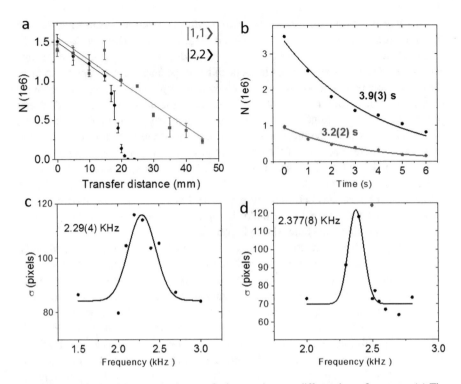

Fig. 9.8 Transferring Rb atoms in the transfer beam using two different hyperfine states. (**a**) The number of Rb atoms as a function of the transfer distance for the $|2, 2\rangle$ state and the $|1, 1\rangle$ state, showing that the sharp loss at 2 cm is a magnetic effect. (**b**) The lifetimes of Rb in $|2, 2\rangle$ at 0 and 2 cm, which are almost the same. (**c**) Parametric measurement of the trap frequency at 0 cm. (**d**) Parametric measurement of the trap frequency at 2 cm

instead of $|2, 2\rangle$, as shown in Fig. 9.8a. Part of the reason it was difficult to identify this problem is that every measurement we could make looked perfectly normal at this position of 2 cm where we lost all the atoms. Figure 9.8b–d shows that the lifetime and the transverse trapping frequency is exactly the same at this position as it is in the initial position of the quadrupole trap. Note that the trap lifetime here is relatively short because we were using a very deep trap in an effort to combat this sharp loss.

These results are perhaps not surprising since a sharp magnetic field gradient would affect the axial trap frequency, which is ≈ 5 Hz, much more than the transverse frequencies. Moreover, the loss of atoms is commensurate with the reduction in the trap depth, and so the temperature of the atoms that remain is such that the ratio of the effective trap depth to the temperature is similar. Therefore, we need other ways to diagnose this problem, which are shown in Fig. 9.9. The first test we would like to do is confirm that the problem is coming from some feature that is specifically at the position of 2 cm. Other possibilities which made it difficult to identify the problem are that the motion profile of the stage (to be discussed later)

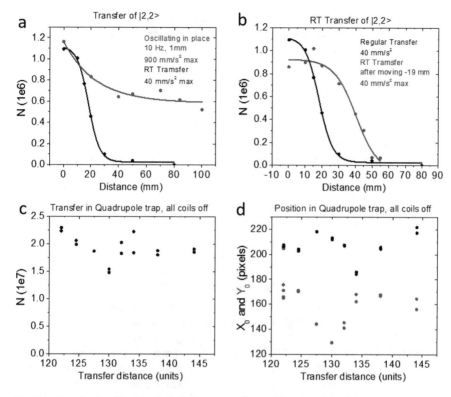

Fig. 9.9 Transferring Rb atoms past the bump. (**a**) The oscillate function of the stage can be used to move the stage a large total distance (red) without it moving more than several mm from the initial position. This shows that the loss is occurring at a specific position (black) rather than a specific distance. (**b**) We can also initially move backwards by 2 cm, and then move forwards with the same profile (red). This shows that the sharp loss occurs 2 cm further than the normal profile (black), which suggests again that the loss occurs at a specific position. (**c**) Moving the quadrupole trap further, we see loss when we stop at exactly the position of the magnetic field bump. (**d**) The horizontal (clack) and vertical (red) position of the cloud as we move past the bump position. This shows that there is a sharp change of the field at the same position where we lose atoms from the transfer beam

could cause a lot of loss from too much acceleration, and it happens to be at 2 cm; or that the transfer beam could be clipping on something in the science cell when the focus is between 0 and 2 cm.

To address this question, we can use the oscillate function of the stage to move the stage back and forth several mm at a certain frequency, thus giving a total distance we can vary without moving very far. This is shown in Fig. 9.9a, where the loss as a function of transfer distance from the oscillation is much more shallow than when moving linearly in the direction of the science cell. A more direct measurement is shown in Fig. 9.9b, where we can first move backwards 2 cm, and then we see that the sharp loss occurs after 4 cm, which is exactly in the same position. Thus, it is now very clear that there is a sharp yet localized magnetic field gradient at 2 cm.

9.3.4   Magnetization in the Chamber

There is nothing we can do to mitigate this problem, even when using a very deep transfer beam. The gradient is just too strong to overcome with an optical trap whose axial frequency is 5 Hz. Moreover, K is lighter and thus more easily affected by such perturbations. It is also very difficult to use less magnetic states of K, and we are limited by the stable combinations of K and Rb, both of which are stretched in the $F = 9/2$ manifold for K. We tried to use several coils oriented along the transfer direction to dynamically shim this gradient in such a way to cancel it out, but it is just too sharp and too localized, such that the profile from any coil cannot compensates it adequately.

It is important to understand what this sharp magnetic field gradient is from. As was discussed in Chap. 8, the small cube in which this plugged-quadrupole evaporation is performed houses the electrodes where they turn 90° to connect the science cell with the large cube which holds the feedthroughs. Additionally, the small cube is where we transfer from the tungsten rods in the science cell to the SS316 that continues all the way to the feedthroughs. The tungsten rods are connected to the SS316 rods using SS316 hypodermic tubing which is crimped onto both rods, and a slit is cut into the gap between the two rods to avoid virtual leaks. These junctions between the right-angle SS316 sections (see Chap. 8) and the tungsten rods are centered very nearly at 2 cm, and our best guess is that they cause the sharp magnetic field gradient.

It is well known that mechanical stress can induce magnetic domain formation in otherwise minimally magnetic materials like SS316. The SS316 rods have been rounded, and the crimped and machined SS316 hypodermic tubing sections certainly have substantial mechanical stress. Further, the gradient field from the quadrupole field causes fields of ≈ 100 G at this location, which are sufficiently high to eventually magnetize a magnetically susceptible material. We tried to de-Gauss the magnetic domains by reversing the field direction, and by leaving the magnetic field on when we move the coils back to the MOT chamber. Typically, we turn the current off before we return the coils to the MOT chamber. None of these approaches were successful in completely removing the magnetization, but the axial dipole force in the transfer beam is very small and thus optical transfer is very unforgiving.

9.3.5 Moving Further in the Magnetic Trap

Other groups have dealt with magnetization issues during transfer (magnetic rare-earth atoms are especially susceptible to this issue). Our solution was to use the large magnetic field gradient from the quadrupole trap to completely swamp the local, sharp gradient from the magnetized region. There are two options: either we move the quadrupole trap and optical trap together over the first several cm in a magnetic-optical hybrid trap [14, 21], or we initially transfer further in the magnetic trap and evaporate several cm past the center of the small cube. Since the plug beam is already aligned through the science cell, we decided to pursue the second option and all we had to change was the focus position of the plug by several cm (the cart coils had to be modified slightly to accommodate obstacles which arose within the additional several cm, but this was quite straightforward).

Therefore, we can now magnetically transfer up to 5 cm further than the center of the small cube, and by transferring to variable positions within this 5 cm, we can see if there is a significant change in the magnetic field. Any change in the background field will move the position of the zero-field region of the quadrupole trap and in turn move the cloud. This is shown in Fig. 9.9c, d, where the number is adversely affected by transferring to exactly the location of this sharp magnetic field bump. Transferring past it with sufficient speed appears to mitigate any loss, and the heating that results is minimal. Figure 9.9d shows how the cloud position changes near the location the sharp gradient, and this measurement corroborates all the measurements with the optical transfer. Therefore, we solved this problem by transferring $\approx$4 cm further in the cart quadrupole coils and performing evaporative cooling offset from the center of the cube. We load the transfer beam from this position as discussed above.

9.3.6 The Optical Transfer Efficiency

The transfer efficiency is now routinely $\approx$90% for both species to go the remaining $\approx$12 cm. The transfer profile is an s-curve to maintain constant jerk, and reaches a maximum acceleration of $\approx$480 mm/s^2. The motion profile of the stage was carefully chosen based on the axial trap frequency of the transfer beam [3, 10], which is $\approx$6 Hz for Rb and $\approx$ 8 Hz for K, though it is slightly non-adiabatic [4] and some heating is observed. The acceleration, velocity, and position during the motion of the stage is shown in Fig. 9.12a. We find that the temperature after transfer is slightly higher than before transfer, and we find that the difference depends on the initial temperature, suggesting that the efficiency is limited by the trap depth when the initial cloud is too hot. This will be discussed more in the next section. The clouds are typically 20 μK for both K and Rb after transfer to the science cell.

Figure 9.10 shows how we mount the optics for the transfer beam, and the two mirrors on the stage are mounted together at exactly a right angle using a custom

Fig. 9.10 The optical layout of the transfer beams. All the optics are stainless steel, except the right-angle mirror mount on the stage which is aluminum

aluminum frame mounted onto the stage. These mirrors are not adjustable, so it is quite important that the incoming beam is parallel to the axis of translation of the stage. We do this using a pinhole on the stage and the first two mirrors after the fiber, while the intermediate lenses are removed. The springs on typical kinematic mounts can cause substantial pointing noise resonances, and these resonances can be excited by the acceleration (and any minimal jerk) of the transfer beam during its motion. We find that using the corner-cube-style aluminum block instead of kinematic mirror mounts slightly reduces the heating of the clouds during transfer. Note that we observe a finite jerk at the very beginning of the motion profile using the digital scope program for the stage. We can vary the parameters for the stage feedback and motion, but they have been observed to not significantly affect the transfer efficiency or transfer heating.

Further, the alignment of the lenses is critical to mitigate any aberrations on the beam. This is particularly important since the axial profile of the beam does the trapping against the acceleration of the motion profile, and thus any astigmatism or coma will adversely affect the transfer efficiency. We align the lenses in the $6.7\times$ telescope and the final focusing lens (the three optics before the stage) one at a time. All three are achromatic doublets. With the beam from the prior two mirrors perfectly aligned through the pinhole on the stage, we place the second lens $f = 200$ mm in the mirror mount, and position the lens such that it is still aligned to the pinhole which should correspond to the exact center of the lens. We can then replace the lens with a mirror to reflect the beam, and we align the kinematic mount to retroreflect the beam back into the fiber. We may have to integrate this process a few times. Then, we do the same procedure with the first lens $f = 30$ mm. Now, the beam on the pinhole has been magnified by $6.7\times$. Finally, we repeat this procedure with the third lens $f = 750$ nm, which is the focusing lens onto the atoms. This procedure produces a perfectly round beam over a large range around the focus.

9.3.7   Heating During Transfer

To characterize the amount of heating during the optical transfer, we move the stage an appropriate distance to carry the atoms into the science cell, and then we move it back to the position of the magnetic quadrupole trap. This round-trip approach allows us to remove any systematics from different imaging systems in the two different regions of the chamber. Figure 9.11 shows the measured temperature and atom number with round-trip transfer and with no transfer at all. The hold time in the transfer beam is the same for both. This measurement is carried out for two different powers of the transfer beam, corresponding to trap depths of 180(10) and 250(10) μK.

We control the temperature of the atoms loaded into the transfer beam by varying the evaporation cut in the plugged quadrupole magnetic trap. The abscissa of this plot is the cut frequency in the magnetic trap, and it is proportional to the temperature. As a point of reference, a cut frequency of 6838 MHz corresponds to a temperature of ≈ 10 μK in the magnetic trap. The difference in temperature after loading into the transfer beam comes from the adiabatic compression due to the higher geometric-mean trap frequency in the transfer beam, and also the different power law of the linear magnetic quadrupole trap versus the harmonic optical dipole trap.

Figure 9.11 shows that the round-trip transfer efficiency is excellent for all cases. However, the temperature after round-trip transfer is always the same regardless of the initial temperature. In all cases, the cloud temperature saturates at $\approx 1/10$th of the trap depth in the transfer beam. This is a common η-parameter for optical traps and optical evaporation. This measurement suggests that there is a very large amount of heating due to the transfer. In fact, we find that virtually all of this heating comes from the very first ≈ 1 mm of transfer, where clearly something non-adiabatic is happening to the cloud at the beginning of the motion profile.

We attribute this to the fact that we are using a magnetic mechanical-bearing stage, which are known to have more friction than air-bearing stages. Indeed, upon monitoring the motion profile of the stage we see that the position error, velocity error, and driver current error all have a spike right as the motion profile initiates. This is needed to overcome the static friction of the system. By varying the servo parameters of the stage, we were able to reduce this spike by a factor of two or so, but we were never able to remove it. Such changes had no observable effect on the heating during transfer, and it seems that these spikes need to be completely removed in order to reduce this heating. We have recently purchased an air-bearing stage (Aerotech ABL10100-LT) with the intention of switching to it from this magnetic mechanical-bearing stage (Aerotech ANT130-110-L-PLUS) that we have been using for all of this transfer beam discussion.

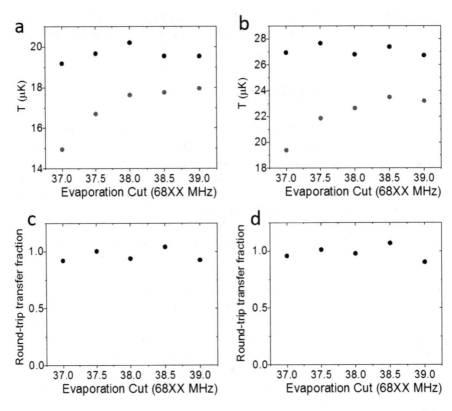

Fig. 9.11 Heating during transfer in the transfer beam. (**a**) The temperature of Rb with (black) and without (red) round-trip transfer in the transfer beam with a trap depth of 180 μK as a function of the evaporation cut in the magnetic trap, which changes the temperature of the cloud. (**b**) The same measurement except with a trap depth of 250 μK. (**c**) The number of atoms with transfer normalized to the number without as a function of the evaporation cut in the magnetic trap, for a trap depth of 180 μK. (**d**) The same measurement except with a trap depth of 250 μK in the transfer beam

9.3.8 An Elliptical Transfer Beam

We recently changed the beam shape of the transfer beam from round with waist ≈45 μm to an elliptical beam of sizes ≈130 μm in the horizontal direction and ≈30 μm in the vertical direction. This serves to increase the axial trap frequency by ~30% and decrease the axial size similarly. We thought this would improve the evaporation efficiency, particularly of K, in the subsequent crossed optical dipole trap, and we thought that this geometry could improve the loading efficiency into the crossed dipole trap. We found that the loading efficiency and the evaporation efficiency were very similar between the round and elliptical transfer beams. The loading efficiency into the transfer beam and the transfer efficiency to the science cell are similar between the two, but the transfer in the elliptical beam may actually have slightly less heating with the same efficiency.

9.4 Optical Evaporation to a Degenerate Mixture—with the Transfer Beam

Once the atoms reach the science cell, we load them into a crossed optical dipole trap composed of: (1) a horizontal optical dipole trap (H OT) with $1/e^2$ radius of $145(10)\,\mu$m in the transfer direction and $30(3)\,\mu$m in the vertical direction, and (2) a vertical beam (V OT) propagating through the ITO-coated plate which is round with radius $100(10)\,\mu$m. The powers of these two beams are $6.5(5)$ W and $4.0(5)$ W, respectively. Figure 9.12b shows the timing diagrams throughout the experiment for K and Rb. The next step is to optically evaporate in the crossed optical dipole trap by reducing the power. We load to the crossed optical dipole trap in 200 ms, while simultaneously reducing the transfer beam power to ~30% of the initial value. The loading efficiency to this trap is ~60% for both K and Rb. Typical conditions in the crossed dipole trap are $N_{Rb} = 9.0(5) \times 10^6$ and $N_K = 1.5(3) \times 10^6$. Many ramping and loading procedures have been used, and most trajectories give essentially the final conditions after evaporation, to be discussed later.

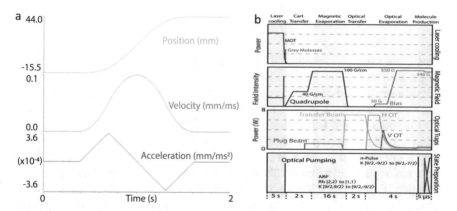

Fig. 9.12 Timing diagrams for the procedure of the experiment. (**a**) The motion profile of stage, showing the position (top), velocity (middle), and acceleration (bottom) commands of the stage as a function of time. Analysis of the position and velocity errors during the motion has been very useful, but is not shown here. (**b**) A timing diagram for the experiment, showing MOT beam powers, magnetic field strengths, and optical trap powers

9.4.1 Compressing the Cloud in the Transfer Beam with a B-Field Gradient

We find that the loading efficiency from the transfer beam into the crossed dipole trap is enhanced by using a magnetic field gradient of ≈25 G/cm on the strong axis, whose center is ≈1 mm above the cloud. This serves to compress the cloud along the axial direction of the transfer beam, reducing the size by ≈30%. However, this effect is less for the elliptical transfer beam because the axial trap frequency is higher and thus the axial size is smaller. After the cloud reaches the science cell, we ramp up this gradient in 200 ms, and then load the crossed dipole trap in 200 ms. We ramp the gradient to zero quickly during the subsequent evaporative cooling in the crossed dipole trap. We have also performed rf-knife pre-evaporation of Rb in the transfer beam plus magnetic field gradient (which we call the squeezed transfer beam) by driving the $|2, 2\rangle$ to $|1, 1\rangle$ in a decreasing frequency sweep that cuts into the cloud from its edge along the axial direction. This evaporation allows us to cool the cloud in the squeezed trap from ≈25 μK to <10 μK while losing a factor of roughly 2 in Rb and almost none in K. This technique yields lower numbers in the crossed dipole trap, but the initial temperature is lower, and thus we end up with slightly more K at the end of evaporation. This method was considered in an effort to preserve our K number, which has been a major limitation, as discussed next.

9.4.2 Loss of K During Evaporation

Again, we sympathetically cool K as Rb is lost first from the trap due to its larger mass and similar polarizability. A major problem, however, is that at higher temperatures when the trap is very deep the effect of gravity is negligible. Therefore, since the polarizability of K is lower than that of Rb by ≈20% at $\lambda = 1064$ nm, we lose both K and Rb during evaporation. Once the trap depth is sufficiently low, the effect of gravity becomes significant, and the trap depth of Rb begins to drop faster than for K. Moreover, the crossover value of the trap depth between these two regimes depends on the vertical waist of the horizontal beam (the waist opposing gravity), since this dipole force directly counters the linear potential from gravity. For $w_z = 45$ μm, the crossover occurs at $T \approx 2$ μK at which the trap depth is $V \approx 20$ μK. For $w_z = 30$ μm the crossover occurs at an even lower trap depth. Since we start with $T = 20(2)$ μK, we lose ≈80% of the initial K before we reach the crossover into efficient sympathetic cooling after which we lose another factor of two or so.

9.4.3 Implementing the 790 nm Beam for Additional Trapping of K

We have tried many approaches to combat this issue. One of which is to load a colder cloud in the crossed dipole trap by pre-evaporative cooling using the rf-knife of Rb in the squeezed transfer beam. While this helped us load a colder cloud into the crossed dipole trap, it is difficult to get it cold enough in the transfer beam such that we can efficiently load into the crossed dipole trap with a trap depth sufficiently low that gravity causes the trap depth of K to be larger than Rb, which we calculate to be $\approx 30\,\mu$K for our trap geometry.

To this end, we turned to species-dependent traps that helped us hold K in the trap more effectively than Rb. We use $\lambda = 790$ nm which was discussed in Chaps. 4 and 5, and can be used to trap K only. The caveat here is that the heating rate for Rb is $\approx 1\,\mu$K/s for a power of ≈ 50 mW in a beam the same size as the horizontal dipole trap. This heating should be manageable for Rb during the first third of evaporation, which is only ≈ 500 ms. After this, gravity will cause the trap depth for K to be larger than for Rb. However, we have never seen any significant improvement from using this beam, and the heating of Rb is much higher than expected. We expect that the issue could be due to the overlap between the 1064 and 790 nm beams.

9.4.4 Optimized Trajectory

Ultimately, we do not rely on the 790 nm beam or the squeeze coils to enhance the loading or evaporation since neither approach significantly enhanced the final conditions. The optimized trajectory using the transfer beam, H OT, and V OT is shown in Fig. 9.12b. After evaporation, we end up with $N_{Rb} = 1.0(2) \times 10^5$ Rb and $N_K = 1.0(2) \times 10^5$ at $T = 400(30)$ nK. These conditions are comparable to the conditions in the old apparatus, and we believe that we can still make further improvements. Figure 9.13a shows time of flight expansion of a quasi-pure BEC of 5×10^4 atoms. Figure 9.13b shows a degenerate Fermi gas of 5×10^4 K atoms with temperature $0.5T_F$. We measure the temperature by fitting the wings of the density profile in expansion to a Gaussian, and extracting the temperature from that fit [7], as discussed in Chap. 6. Fitting the entire cloud to a Fermi–Dirac distribution [7] gives a consistent temperature.

9.5 Eliminating the Optical Transfer Step

The previous sections illustrated the enormous number of complications we encountered using an optical dipole trap with translating focus to carry the atoms through the vacuum chamber from the evaporation region to the science cell. The primary

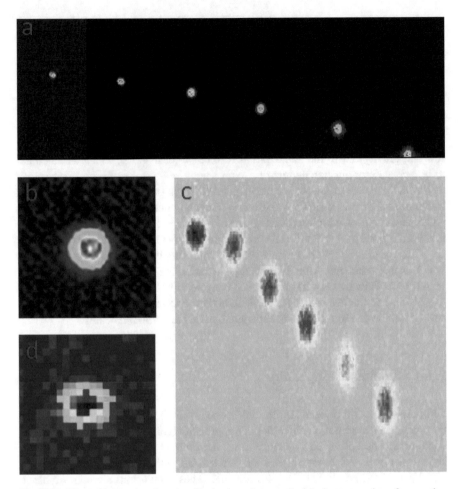

Fig. 9.13 Ultracold atoms and molecules in the science cell. (**a**) shows a series of expansion images of an Rb BEC of 5×10^4 atoms. (**b**) shows an image of 5×10^4 K in expansion, whose cross section fits to a Fermi–Dirac distribution give a temperature of $0.5(1)T_F$. (**c**) shows state preparation of K at 30 G, indicating excellent control of the populations of individual spin states in the $F = 9/2$ manifold. Shown from left to right are $m_F = 9/2, -1/2, -3/2, -5/2, -7/2$, and $-9/2$. The gradient does not allow for separation of the $+m_F$ states. (**d**) shows an image of $<10^4$ Feshbach molecules in expansion

reason to use this transfer is because we have the high-NA microscope objective below the science cell which obstructs the access of the moving coils carrying the atoms from the MOT chamber to the evaporation chamber. If this microscope objective were absent or somehow modified, we could simply use the moving coils to carry the atoms from the MOT chamber to the science chamber directly without the need for an intermediate chamber for evaporative cooling.

It turns out that if the mount for the objective and the dichroic filter above it are indeed modified, we could potentially use the cart coils to go all the way to the science chamber. We would just need to change the separation of the moving coils by ≈1 cm, which is a change of only <10%. With the transfer beam eliminated, we can load directly from the magnetic quadrupole trap after evaporation to the crossed optical dipole trap. Thus, we never need to go from a round(-ish) trap to a very elongated trap, and then back to a round trap. We have recently abandoned the optical transfer procedure (without even trying the air-bearing stage), and are moving the quadrupole coils all the way to the science cell. The objective mounting scheme is being modified in parallel such that we can use it for high-resolution vertical imaging.

Care must be taken to avoid clipping the hot cloud in the quadrupole trap on the aperture of the electrodes in the science cell. We find that ≈20–30% of the atoms will be lost during the typical quadrupole transfer scheme. However, by increasing the gradient to the final evaporation gradient before reaching the science cell this loss is mitigated. Additionally, we found that beginning the microwave evaporation before the coils reach the science cell is also effective since microwaves efficiently reach well past the science cell, and the plug beam is not important during the beginning stages of evaporation. Further, we have decreased the size of the plug beam to ≈20 μm, which requires less power. Lastly, we now decrease the gradient of the quadrupole field as the evaporation progresses. This reduces the density and minimizes 3-body loss, which becomes more significant as the cloud becomes colder.

As a result of these improvements, we are able to reach even better conditions than before, except now in the science cell. This already puts us in a much better initial condition for loading the crossed optical dipole trap. Nevertheless, we would like to evaporate to ≈2 μK in the quadrupole trap instead of 16 μK. As stated above, the subsequent optical evaporation becomes much more efficient for K at these lower temperatures where the gravitational sag of Rb is significant. Work to further improve the magnetic evaporation trajectory below 10 μK is ongoing.

A potential downside of this scheme is that far more of the evaporation is done in the science cell, and thus more atoms will fall onto the ITO-coated plate. Beginning evaporation outside of the science cell also helps greatly in this regard. In principle, we would like to use a blue-detuned light sheet beam under the cloud to minimize the flux of atoms falling onto the bottom plate. Moreover, we have begun to use UV LEDs in the science cell region, similarly to the MOT region as discussed above. By applying them for a few seconds during every run of the experiment, we minimize the alkali pressure in the vicinity of the plates. We observe that the amount of alkali atoms in the science cell region is small enough that no increase in the pressure is observed from using the UV LEDs, as we expect.

At the time of this writing, optical trapping with this new scheme has not yet been tested. All the following discussion of state preparation, production of molecules, and the new physics discussed in the next chapter has been done with the transfer beam scheme discussed above in depth. We expect significant improvements to result from this new scheme without the transfer beam, particularly in the K conditions at the end of evaporation and the number of molecules we can produce.

9.6 State Preparation, and Producing Molecules

During the optical evaporation, we transfer K and Rb from $|9/2, 9/2\rangle$ and $|2, 2\rangle$ to the Feshbach-resonant states $|9/2, -9/2\rangle$ and $|1, 1\rangle$ [23]. This is done at 26 G using Adiabatic Rapid Passages (ARPs) by sweeping the frequencies. For Rb, we sweep the frequency from 6.898492 to 6.898992 GHz in 2 ms with a power to the antenna of 40(2) dBm. For K, we sweep the frequency from 10.2 to 8.5 MHz in 20 ms with a power to the rf coil of 35(2) dBm. We demonstrate excellent control of the population in all ten spin states for K in Fig. 9.13c by during Stern–Gerloch spin separation during expansion [20].

We initially had a difficult time coupling rf and microwaves into the science cell, and we eventually learned that the electrodes are partially the reason for this. As discussed in Chap. 7, they act as an effective Faraday cage for electric fields, and thus they attenuate the rf and microwave signals. In order to have a good enough coupling between the rf and microwave coils and the atoms in order to reach Rabi frequencies of $\Omega \approx 10$ kHz, we must use >30 dBm at 10 and 80 MHz and >40 dBm at 6.8 and 8 GHz.

The disadvantage of using large rf fields is that they can couple onto electronics, such as the servo for the magnetic field, and rectify to DC or frequencies below the servo bandwidth of ≈ 1 kHz. To address this issue, we used a lot of filtering on the Hall probe for the magnetic field servo, and on the servo input. We found that the grounding scheme of the B-field coils and their servos was a very important part of removing this coupling. We also needed to shield many of the electronics to minimize their exposure to the rf field. In the end, we were able to do this sufficiently well that the rf has negligible effects on the magnetic field, but it took us some time to learn how to do this successfully.

Once both species are in the Feshbach resonant states, we sweep the magnetic field up to 560 G in 500 ms towards the end of evaporation such that the atoms are above the Feshbach resonance as the final temperature is reached. Then, by sweeping the field from above to below the resonance we can associate K and Rb into Feshbach molecules [5, 17, 23]. Feshbach molecules can be detected by probing with resonant light of either species (Fig. 9.13d shows $<10^4$ Feshbach molecules imaged using K). The final field used after the Feshbach sweep imbues the molecules with a binding energy of ~ 100 kHz, and so light resonant with either atom dissociates the molecules [23]. However, they can be differentiated from free K atoms by transferring free K to the $|9/2, -7/2\rangle$ state. The binding energy can be sufficiently large compared to the Rabi frequency of this π-pulse in order to not affect the Feshbach molecules. This pulse is at 80 MHz, and the time of the π-pulse is typically $\sim 100 \,\mu$s.

To make ground-state molecules, all unpaired atoms can immediately be removed with resonant light as soon as the molecules are transferred to the ground state, and therefore no state transfer manipulation is needed. The STIRAP pulse time is typically $\sim 10\,\mu s$. The STIRAP beams propagate down along the vertical direction. It is important that they are along the quantization (B-field) axis, since their polarization is important [18]. In both the old apparatus and now the highest molecular yield from a bulk gas was found using atomic temperatures near T_C and T_F of ~ 100–$200\,nK$. In such cases, we are able to produce $\approx 1 \times 10^4$ ground state molecules with a temperature of ≈ 2–$3\,T_F$, as shown in Fig. 9.13d. Such conditions are a factor of three worse than in the old apparatus [18], and our work on this issue is ongoing. We believe the new motion stage for the transfer beam or removing optical transfer all together will help us with this problem.

9.7 Optical Lattices

The horizontal lattice beams for the 3D optical lattice will come through the diagonal windows of the octagonal cell, and they are the only beams which use these paths. This allows us to keep the horizontal lattice beams simple. They are retro-reflected with a $\lambda/4$, lens, and mirror after the incident beam exits the science cell. The lens collimates the beam after the incident pass, and refocuses the retro-reflection back to the same place in the center of the science cell. We align the retro-reflection using a kinematic mount for the mirror, and optimize the power that goes back through the optical fiber.

We choose the lattice beam size to be $\omega_z = 40(5)\,\mu m$ in the vertical direction and $\omega_r = 160(10)\,\mu m$ in the horizontal direction. We maintain an angle of $\approx 3^o$ between the horizontal lattice beams and the axes normal to the windows of the science cell through which they pass. This is to avoid superlattice effects due to the $\approx 1\%$ reflection per surface from the cell windows, as discussed in Chap. 4.

We put a $\lambda/2$ on the incident path of the lattice such that we can tune the polarization of the lattice beams. This is important to achieve the magic polarization angle discussed in Chap. 4. The DC electric field will nominally be oriented in the vertical direction, which is also the direction of the magnetic field. Thus, vertical polarization will give a $0°$ angle, and horizontal polarization will give a $90°$ angle. However, since these fields are oriented along the vertical direction, the vertical lattice will always be at a $90°$ angle, and thus we will not be able to reach the magic angle of $\sim 50°$ with this beam. For this reason, the "magic electric field" discussed in Chap. 10 will be very important [12].

9.7.1   Vertical Lattice Path

The vertical lattice path is significantly more complicated, since it has to propagate through all the optical substrates in the science cell. As a result of the baking process, the reflection of the ITO-coated plates is a few percent, and thus we cannot directly form a lattice with normal incidence. Moreover, since the dichroic below the science cell to reflect $\lambda = 1064$ nm cannot be angled by more than a few degrees, we were forced to use a more complicated beam path for the vertical lattice. This path is shown in Fig. 9.14, and it is based on two beams with a $\approx 10°$ with respect to the cell.

The beam that hits the cloud before reflecting off of the dichroic (red) is the vertical optical trap beam discussed above. This beam is already part of the crossed optical trap, and to turn on the vertical lattice we ramp on the beam which is aligned to the cloud after a reflection from the dichroic (green). We use a waist of $100\,\mu$m for both of these beams, but they can be independently changed. The lattice spacing of the optical lattice scales as $(\lambda/2)/\cos(\theta)$, and for $\theta = 10°$, the lattice spacing is 540 nm. The lattice oscillation frequency $\omega_{\mathrm{latt}}^{\theta}$ scales as $\omega_{\mathrm{latt}}^{0}\cos(\theta)$, and thus the oscillation frequency from this configuration is 98% of what it would be with a retroreflected lattice.

This lattice is much easier to align than we initially expected. Both beams are essentially the same, except the position of their focus is different. We can independently align them onto the cloud. Both beams can be aligned such that either the incoming or the reflected beam hits the cloud. We can identify which is which by reducing the transfer beam to allow the atoms to slide down the vertical beam. By observing whether the atoms slide down sloping left or right, we are immediately able to identify whether the beam is hitting the atoms before or after bouncing off the dichroic. Figure 9.14b shows horizontal images of the coherent diffraction (see Chap. 4) of an Rb BEC off of the vertical lattice potential when it is pulsed on for a few μs.

An issue of immense important for spectroscopically selecting a single layer in the vertical lattice is that each antinode moves during the procedure by an amount which is small compared to their separation $\approx 0.5\,\mu$m. In order to guarantee the phase stability between the two beams, we interfere them both on a beam splitter. As shown in Fig. 9.14, the lattice is formed by only one of the beams reflecting off of the dichroic and thus we are particularly sensitive to its drift. If we interfere the two beams after they both reflect off the dichroic, then the drift of the dichroic would be common mode. Therefore, we interfere the incoming vertical optical trap beam before the cell with the reflection of the vertical lattice beam after the cell and the dichroic.

This allows us to maintain a constant phase of the lattice when only one beam goes through the dichroic. We use a ring piezo-actuator on the holder of the fiber for the vertical lattice so that we can change its phase independently. We use the interference fringe on a photodiode to feed back to this piezo. All other optics in the setup of these two beams are stainless steel mounts which were arranged in the

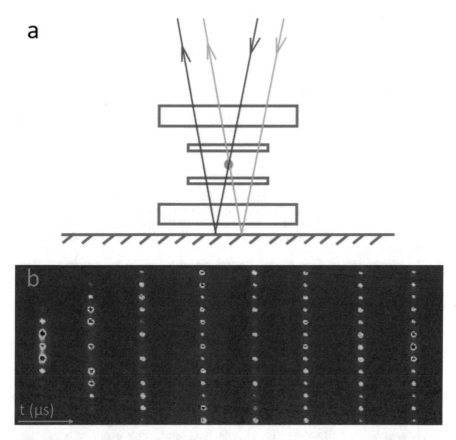

Fig. 9.14 The vertical lattice. (**a**) Scheme for forming the vertical lattice using a reflection from the dichroic below the science cell. All the surfaces in blue are on substrates in the cell such as its vacuum windows and the ITO-coated plates inside the cell. The hashed blue surface is dichroic below the science cell which reflects $\lambda = 1064\,\text{nm}$. The vertical optical dipole trap hits the cloud at a $\approx 10°$ angle before hitting the dichroic (red). The vertical lattice beam has a similar angle, but it hits the dichroic first (green) and interferes with the incoming beam after reflecting off of the dichroic. (**b**) Horizontal imaging of the diffraction of an Rb BEC off of the vertical lattice. The images from left to right show the lattice pulsed on for successively longer times, from $1\,\mu\text{s}$ to $8\,\mu\text{s}$, with 8 ms expansion time. Notice that the central component is depleted and re-emerges during this time. Note that a tilt angle of a few degrees has been removed, which is due to an angle between the horizontal imaging system and the vertical direction

most stable way possible in an effort to ensure that the drift of the relative phase is predominantly caused by the dichroic.

References

1. D.M. Brink, C.V. Sukumar, Majorana spin-flip transitions in a magnetic trap. Phys. Rev. A **74**, 035401 (2006)
2. A. Burchianti, G. Valtolina, J.A. Seman, E. Pace, M. De Pas, M. Inguscio, M. Zaccanti, G. Roati, Efficient all-optical production of large ^{6}Li quantum gases using D_1 gray-molasses cooling. Phys. Rev. A **90**, 043408 (2014)
3. A.P. Chikkatur, Y. Shin, A.E. Leanhardt, D. Kielpinski, E. Tsikata, T.L. Gustavson, D.E. Pritchard, W. Ketterle, A continuous source of Bose-Einstein condensed atoms. Science **296**, 2193–2195 (2002)
4. A. Couvert, T. Kawalec, G. Reinaudi, D. Gury-Odelin, Optimal transport of ultracold atoms in the non-adiabatic regime. Europhys. Lett. **83**(1), 13001 (2008)
5. J.P. Covey, S.A. Moses, M. Garttner, A. Safavi-Naini, M.T. Miecnkowski, Z. Fu, J. Schachen-mayer, P.S. Julienne, A.M. Rey, D.S. Jin, J. Ye, Doublon dynamics and polar molecule production in an optical lattice. Nat. Commun. **7**, 11279 (2016)
6. K.B. Davis, M.O. Mewes, M.R. Andrews, N.J. van Druten, D.S. Durfee, D.M. Kurn, W. Ketterle, Bose-Einstein condensation in a gas of sodium atoms. Phys. Rev. Lett. **75**, 3969–3973 (1995)
7. B. DeMarco, D.S. Jin, Onset of fermi degeneracy in a trapped atomic gas. Science **285**(5434), 1703–1706 (1999)
8. R. Dubessy, K. Merloti, L. Longchambon, P.-E. Pottie, T. Liennard, A. Perrin, V. Lorent, H. Perrin, Rubidium-87 Bose-Einstein condensate in an optically plugged quadrupole trap. Phys. Rev. A **85**, 013643 (2012)
9. D.R. Fernandes, F. Sievers, N. Kretzschmar, S. Wu, C. Salomon, F. Chevy. Sub-doppler laser cooling of fermionic 40 k atoms in three-dimensional gray optical molasses. Europhys. Lett. **100**(6), 63001 (2012)
10. T.L. Gustavson, A.P. Chikkatur, A.E. Leanhardt, A. Görlitz, S. Gupta, D.E. Pritchard, W. Ketterle, Transport of Bose-Einstein condensates with optical tweezers. Phys. Rev. Lett. **88**, 020401 (2001)
11. M.-S. Heo, J.-Y. Choi, Y.-il Shin, Fast production of large ^{23}Na Bose-Einstein condensates in an optically plugged magnetic quadrupole trap. Phys. Rev. A **83**, 013622 (2011)
12. S. Kotochigova, D. DeMille, Electric-field-dependent dynamic polarizability and state-insensitive conditions for optical trapping of diatomic polar molecules. Phys. Rev. A **82**, 063421 (2010)
13. P.D. Lett, R.N. Watts, C.I. Westbrook, W.D. Phillips, P.L. Gould, H.J. Metcalf, Observation of atoms laser cooled below the doppler limit. Phys. Rev. Lett. **61**, 169–172 (1988)
14. Y. Lin, R.L. Compton, K. Jimenez-Gercia, J.V. Porto, I.B. Spielman, Synthetic magnetic fields for ultracold neutral atoms. Nature **462**, 628–632 (2009)
15. M. Meucci, E. Mariotti, P. Bicchi, C. Marinelli, L. Moi, Light-induced atom desorption. Europhys. Lett. **25**(9), 639 (1994)
16. G. Modugno, C. Benkő, P. Hannaford, G. Roati, M. Inguscio, Sub-doppler laser cooling of fermionic ^{40}K atoms. Phys. Rev. A **60**, R3373–R3376 (1999)
17. S.A. Moses, J.P. Covey, M.T. Miecnikowski, B. Yan, B. Gadway, J. Ye, D.S. Jin, Creation of a low-entropy quantum gas of polar molecules in an optical lattice. Science **350**(6261), 659–662 (2015)
18. K.-K. Ni, S. Ospelkaus, M.H.G. de Miranda, A. Pe'er, B. Neyenhuis, J.J. Zirbel, S. Kotochigova, P.S. Julienne, D.S. Jin, J. Ye, A high phase-space-density gas of polar molecules. Science **322**(5899), 231–235 (2008)
19. W. Petrich, M.H. Anderson, J.R. Ensher, E.A. Cornell, Behavior of atoms in a compressed magneto-optical trap. J. Opt. Soc. Am. B **11**(8), 1332–1335 (1994)
20. T.M. Roach, H. Abele, M.G. Boshier, H.L. Grossman, K.P. Zetie, E.A. Hinds, Realization of a magnetic mirror for cold atoms. Phys. Rev. Lett. **75**, 629–632 (1995)

21. J.F. Sherson, C. Weitenberg, M. Endres, M. Cheneau, I. Bloch, S. Kuhr, Single-atom-resolved fluorescence imaging of an atomic Mott insulator. Nature **467**, 68–72 (2010)
22. J.J. Zirbel, Ultracold fermionic feshbach molecules. PhD thesis, University of Colorado, Boulder (2008)
23. J.J. Zirbel, K.-K. Ni, S. Ospelkaus, J.P. D'Incao, C.E. Wieman, J. Ye, D.S. Jin, Collisional stability of fermionic feshbach molecules. Phys. Rev. Lett. **100**, 143201 (2008)

Chapter 10
New Physics with the New Apparatus: High Resolution Optical Detection and Large, Stable Electric Fields

With the reasonable conditions of KRb molecules in the new apparatus reported in the last chapter (and further improvements expected), we are ready to start exploiting its novel capabilities. Again, these capabilities include large, versatile electric fields; AC electric field polarization control; and high resolution in situ detection and addressing. I will now discuss our work in the new apparatus which uses these tools. Note that at the time of this writing our work on some of these subjects is very preliminary and limited. Nevertheless, this chapter is intended to describe current research directions, and directions we are preparing to investigate imminently.

10.1 Large, Stable Electric Fields

We can begin to apply large electric fields to molecules confined in a bulk 3D gas or a 1D lattice either along the vertical or horizontal directions. We measure the electric field by performing Stark spectroscopy [12, 20, 23, 26, 28], in which we measure the energy of a rotational transition as a function of the voltages on the electrodes. See Chap. 3 for a discussion of DC Stark shifts. For these measurements we keep the electrodes in the homogeneous configuration discussed in Chap. 7. The frequency of the $N = 0$ to $N = 1$ rotational transition varies between 2.23 GHz at $E = 0$ to ≈ 8 GHz at $E = 30$ kV/cm, and thus the rod electrodes need to have sufficient AC coupling over this entire range (see Chap. 8). A schematic of the electrode geometry with the molecules in optical lattices is shown in Fig. 10.1.

© Springer Nature Switzerland AG 2018   219
J. P. Covey, *Enhanced Optical and Electric Manipulation of a Quantum Gas of KRb Molecules*, Springer Theses, https://doi.org/10.1007/978-3-319-98107-9_10

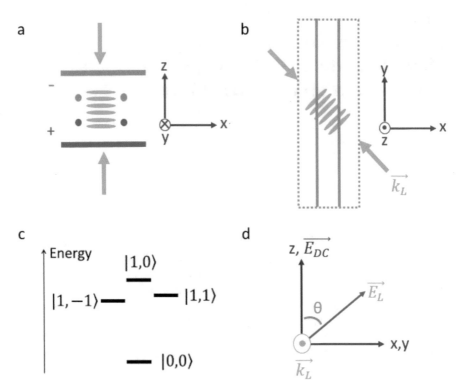

Fig. 10.1 A schematic for applying electric fields to molecules in optical lattices. (**a**) A vertical lattice propagates through the ITO-coated plates. (**b**) The horizontal lattices propagate at a 45° angle with respect to the symmetry axis of the electrodes. (**c**) The $E = 0$ level structure of the $N = 0$ and $N = 1$ rotational states. (**d**) The geometry of the DC electric field and electric field of the lattice beam from one of the horizontal lattices. The angle between the DC electric field and lattice electric field θ is tunable between 0 and $\pi/2$

10.2   Driving Rotational Transitions: AC Electric Field Coupling

As discussed in Chap. 7 the rods were designed with the capability of applying microwave-frequency AC electric fields. The orientation of this electric field can be controlled outside the vacuum chamber, and thus by changing the relative orientation between the AC driving field and the quantization axis (either the DC electric field or the magnetic field) we can choose the polarization of the microwave field. This allows us to selectively drive rotational transitions of a specific relative angular momentum; either a π transition when $E_{DC} \parallel E_{AC}$, or σ^+ and σ^- when $E_{DC} \perp E_{AC}$. Previous work in the old apparatus did not have any polarization control, and thus it was not possible to selectively drive these transitions. Only energy selectivity allowed us to choose the state to which we coupled in the old apparatus.

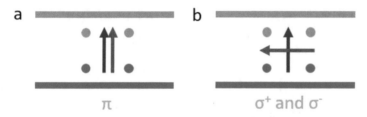

Fig. 10.2 Applying AC electric fields and driving rotational transitions. (**a**) The DC electric field (purple arrow) and the AC electric field (red arrow) are parallel, driving π transitions with $\Delta m_I^{\text{tot}} + \Delta m_N = 0$. (**b**) The DC electric field (purple arrow) and the AC electric field (red arrow) are perpendicular, driving σ^+ and σ^- transitions with $\Delta m_I^{\text{tot}} + \Delta m_N = \pm 1$

Therefore, we expect the orientation of E_{AC} to affect the excited rotational states to which we can couple (see Fig. 10.2). The ground state is $|N, m_N, m_I^K, m_I^{\text{Rb}}\rangle = |0, 0, -4, 1/2\rangle$, where m_I^K and m_I^{Rb} are the nuclear spin projections of K and Rb onto the magnetic field, respectively. Therefore when we drive π transitions with $E_{\text{DC}} \parallel B \parallel E_{\text{AC}}$, the accessible excited states are $|1, 0, -4, 1/2\rangle$, $|1, 0, -3, -1/2\rangle$, $|1, -1, -3, 1/2\rangle$, $|1, -1, -4, 3/2\rangle$, $|1, 1, -4, -1/2\rangle$ such that $m_{\text{tot}} \equiv m_N + m_I^K + m_I^{\text{Rb}}$ remains the same ($m_{\text{tot}} = -7/2$). When $E_{\text{DC}} \parallel B \perp E_{\text{AC}}$, both the σ^+ and σ^- transitions are available, so there are many options such that $m_{\text{tot}} = -9/2$ or $-5/2$.

10.3 "Magic" Electric Fields

The AC polarizability as a function of the angle between the quantization axis and the trapping light polarization axis discussed in Chap. 4 was in the absence of an electric field (see Fig. 4.9). However, when a large electric field is applied it polarizes the molecules thereby mixing opposite parity rotational states, and the dependence on the polarization angle changes [17]. At $E = 0$ the polarizability of the $|1, 0\rangle$ state is higher than $|0, 0\rangle$ for $0°$ and lower for $90°$. Conversely, when the electric field is sufficiently large this trend reverses. At a special field E_{magic} the differential polarizability (difference in polarizability of the two states) is zero for all angles, as shown in Fig. 10.3.

For KRb this occurs at $10\,\text{kV/cm}$ [17], which is coincidentally similar to the field at which J_z is maximized [13]. As stated in the previous chapter this "magic electric field" is particularly important because the vertical lattice is oriented along the electric and magnetic field direction, and thus the relative polarization angle will always be $90°$ (as in Fig. 10.3b). Accordingly, the magic electric field will be the only way to reach zero differential polarizability from the vertical OT beam.

We can study the dependence of the differential polarizability on electric field by loading molecules into a 1D lattice along the horizontal axis such that we can tune the polarization angle relative to the field between 0 and $90°$ (see Fig. 10.1b, d).

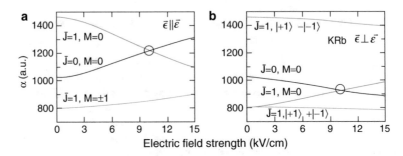

Fig. 10.3 The magic electric field. (**a**) The polarizability of the $\widetilde{N} = 0$ and 1 states (here called $\widetilde{J}$) as a function of the electric field when the trap polarization $\vec{\epsilon}$ is parallel to the electric field $\vec{E}$. (**b**) The polarizability of the $\widetilde{N} = 0$ and 1 states as a function of the electric field when the trap polarization $\vec{\epsilon}$ is perpendicular to the electric field $\vec{E}$. Reproduced from Ref. [17]

We can measure the polarizability of $|\widetilde{0}, 0\rangle$ and $|\widetilde{1}, 0\rangle$ in two ways, such as in Ref. [22] (the tilde denotes the basis of rotational states in the presence of a large electric field where eigenstates are no longer pure states. See Chap. 3). The first method is to measure the lattice depth by performing parametric heating, as discussed in Chap. 4. The second method is to use the transition frequency in the lattice compared to the transition frequency in free space (expansion).

10.4   Towards Evaporative Cooling

With the ability to reach large electric fields in hand, we can now enter the regime of stabilized dipoles in 2D by using the vertical 1D lattice. Previous work in Ref. [8] was only able to reach $d = 0.2$ Debye and $v_z = 23$ kHz, where the ratio of elastic to inelastic collisions is only ≈ 200 at $T = 500$ nK [33]. Under these conditions the lifetime was measured to be ≈ 1 s [8], which is similar to the case of a 3D gas with no electric field [24]. However, with the ability in hand to reach $d = 0.4$ Debye and $v_z = 50$–100 kHz at $T = 200$ nK, the ratio of elastic to inelastic collisions can be >1000 [33]. This is expected to lead to significantly longer lifetimes. In this setting it is possible to perform spin physics between $|\downarrow\rangle = |0, 0\rangle$ and $|\uparrow\rangle = |1, 0\rangle$ without a full 3D lattice, where strongly repulsive dipolar interactions stabilize the system against the s-wave collisions discussed in Chap. 3.

With such strong dipolar repulsion between molecules in a 2D system, we can return to our previous efforts on evaporative cooling as discussed in Chap. 6. We can tilt the 2D traps in the radial direction of the vertical 1D lattice using a horizontal gradient of the electric field as discussed in Chap. 7. Gradients of ≈ 100 V/cm² are sufficiently high to begin spilling the molecules. We work in the homogeneous field case where $V_{\text{rod}}/V_{\text{plate}} = 0.4225$, except that we bias the rod voltages in the horizontal direction as discussed in Chap. 7. The large ratio of elastic to inelastic collisions will significantly improve the thermalization time relative to the chemical reaction time, leading to much more efficient evaporation than in our previous efforts discussed in Chap. 6.

Also as discussed in Chap. 6, we previously started from evaporation with $T/T_F = 3 - 4$ in the 2D pancakes. Our current strategy in the new apparatus will allow us to begin at $T \approx T_F$, which combined with the improved elastic to inelastic collision ratio should enable efficient evaporation to $T < 0.5T_F$. The improvement in the initial PSD in the pancakes comes from two significant changes. The first comes from a change in the aspect ratio of the initial optical dipole trap. Previously the waist of the beams in the vertical direction was $\approx$45 μm, but now we use $\approx$25 μm (discussed in Chap. 9). This reduces the number of pancakes occupied by $\approx$2$\times$, which increases T_F by $\sqrt{2}$. We are also preparing another H OT beam with a vertical waist of $\approx$5–10 μm to compress the cloud into even fewer pancakes. The other improvement will come from cooling the atoms in the 1D lattice prior to making molecules. Previously, we loaded molecules in a 3D gas into the 2D pancakes. Now we plan to cool the atoms in the 2D pancakes, and makes molecules in 2D from sufficiently cold atomic clouds. If necessary, atomic evaporation in 2D can be done using a radial magnetic field gradient [31]. The coils around the science cell discussed in Chaps. 7 and 8 are designed for this purpose. As a result of these improvements we anticipate the initial conditions for molecular evaporation to be $T \approx T_F$ with $N > 10^4$ molecules per pancake.

10.5   High Resolution In Situ Imaging of Polar Molecules

While ground-state molecules can be imaged directly [30], such imaging results in a poor signal-to-noise ratio due to the lack of closed cycling transitions. However, it became clear that molecules can be detected in the same way they are created: by reversing the STIRAP process and the magneto-association process of the Feshbach molecule, a ground-state polar molecule can be converted to a pair of free atoms with $\approx$90–95% fidelity [23, 34], which is limited by the presence of other states that make the three-level lambda-system picture only an approximation [1, 29] (for further background, see, e.g., [6, 21]). It is important to point out, however, that the STIRAP process is state-selective, and energy and angular momentum selection rules allow all excited rotational states to remain decoupled from the STIRAP sequence.

Our initial work on high resolution in situ imaging of polar molecules is based on absorption imaging. Many groups have used high resolution in situ absorption imaging of 2D atomic samples with $R_{\text{min}} \approx 1$ μm for many species. This has been done both in a lattice and in a bulk 2D gas for both bosons and fermions. Species including Cs [11], ^{87}Rb [9], ^{40}K [10], and ^{6}Li [14, 19] have been used. A nice overview of the setup and characterization of such imaging systems can be found in Refs. [3, 15, 16]. As discussed in Chap. 6, we can detect molecules by imaging K or Rb, and high resolution absorption imaging of both has been demonstrated.

The new KRb apparatus has a microscope objective for high resolution in situ detection of polar molecules. Initial experiments will be done using absorption imaging [11], but future directions will be based on Raman sideband cooling (RSC) of K and Rb for operating a dual quantum gas microscope. Details of the new apparatus and its optical capabilities are outlined if Ref. [5]. Future work towards a quantum gas microscope will use an objective with numerical aperture (NA) of 0.6–0.65, which corresponds to an optical resolution at $\lambda = 780$ nm of $R_{min} = 790$–730 nm (Rayleigh criteria). However, the current objective has an NA of 0.53, with a resolution of 900 nm. We use an Andre electron-multiplying CCD (EMCCD) for detection. We currently image with a magnification of $M \approx 24$ such that R_{min} is mapped onto several pixels.

10.6   Single-Molecule Addressing in an Optical Lattice

Sending a beam through the objective backwards allows us to focus a beam onto the atoms, which can impart an AC Stark shift onto that region of the array. For molecules most wavelengths have deleterious effects since their photon scattering rates are high due to the enormous number of rovibronic excited states. We have tried many wavelengths for KRb molecules (see Chap. 4), and found 532, 790, and 808 nm to all have prohibitively large scattering rates. Moreover, we found that the polarizability is small and crosses zero at ∼750 nm. Fortuitously, we know that $\lambda = 1064$ nm has a low scattering rate [4] while still providing a large polarizability [22]. Therefore, we would like to use a 1064 nm beam to selectively Stark shift specially tailored regions of the cloud. An objective with resolution of $R_{min} = 900$ (750) nm at $\lambda = 780$ nm can also be used to focus a Gaussian beam of wavelength $\lambda = 1064$ nm to a $1/e^2$ radius of 860 (700) nm.

We now consider what AC Stark shifts result from an addressing beam of these sizes, and we determine what powers will be required to reach a desirable Stark shift. To uniquely address the molecules within the addressing beam, we require that the AC Stark shift be large compared to the Rabi frequency, which is typically ∼10 kHz for such state preparation. Therefore, for a beam of $1/e^2$ radius 860 (700) nm, a power of 10 (7) μW is needed for a 50 kHz differential AC Stark shift of the $|0, 0⟩ - |1, 0⟩$ transition at the center of the beam. To create a larger impurity region as shown in Fig. 10.4, we want an addressing beam with radius of, i.e., 2 μm, for which 50 μW is needed. Note that similar powers are needed for the $|1, -1⟩$. All of these powers are easily manageable. Also, note that the polarization of the addressing beam is linear and perpendicular to the quantization axis, and the AC Stark shift was calculated accordingly [22].

Another exciting possibility for creating an initial condition that will lead to novel out-of-equilibrium many-body dynamics is to have the left half as $| \downarrow⟩$ and the right half as $| \uparrow⟩$, with a hard domain wall in between, similar to recent work with atoms tunneling in a lattice [2]. While this state could be created using AC Stark shifts from an addressing beam, an alternative approach may be to use an electric field

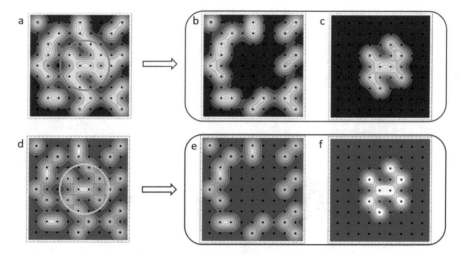

Fig. 10.4 Simulations of atoms on sites of the lattice imaged with two different resolutions. (**a**) Atoms imaged with a resolution of $R_{min} = 900$ nm. The orange circle indicates the region illuminated with a relatively large addressing beam. (**b**) The atoms that remain in one spin state after using the addressing beam imaged with resolution of $R_{min} = 900$ nm. (**c**) The atoms in the other spin state imaged with resolution of $R_{min} = 900$ nm. (**d**) Atoms imaged with a resolution of $R_{min} = 750$ nm. The light-blue circle indicates the region illuminated with a relatively large addressing beam. (**e**) The atoms that remain in one spin state after using the addressing beam imaged with resolution of $R_{min} = 750$ nm. (**f**) The atoms in the other spin state imaged with resolution of $R_{min} = 750$ nm. Reproduced from Reference [7]

gradient across the layer using the electrodes in the science cell of the new KRb apparatus, as outlined in Ref. [5]. Operationally, this approach would be similar to the technique of selecting a single layer as outlined above and in Ref. [5], except that the frequency would be swept in order to flip the spins of all molecules on one side of the domain wall, as in an Adiabatic Rapid Passage pulse, or ARP. Such ARPs are very robust, and any fluctuation in the field would only change the location of the domain wall.

10.7   Holographic Projection of Arbitrary Potentials Using a DMD

We can go beyond simply using a single tightly focused beam, and introduce an array of spots imaged onto the molecules by holographically projecting a spatial light modulator (SLM), such as a digital micro-mirror device (DMD). These techniques have been used to create arbitrary potentials by several groups [2, 18, 27]. In our case we can create arbitrary, random patterns with $\lambda = 1064$ nm which create a disordered landscape as a setting for investigating many-body localization [32].

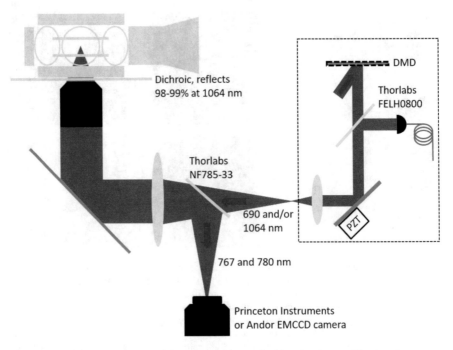

Fig. 10.5 A Schematic for combining the imaging and addressing beams. The imaging wavelengths are 767 and 780 nm, and the addressing wavelengths are 690 and 1064 nm. These can be separated using dichroic filters

The scattering rate at $\lambda = 690$ nm (STIRAP down-leg wavelength) may be too high to use it for this purpose, but since 690 nm is blue-detuned for K and Rb this technique could be used to project an anti-confining potential to reduce the harmonic confinement in the 1064 nm lattice, and thus accommodate a larger $N = 1$ shell, as discussed in Chaps. 4 and 6.

Figure 10.5 shows how the imaging path, single-spin addressing path, and the DMD path are all combined together using dichroic filters. We use a mirror with a piezo actuator to position the addressing beam or DMD onto the cloud. The DMD is the Texas Instruments Digital Light Processing (DLP) LightCrafter 6500 model, with DLPC900 DMD digital controller and 1080p resolution (1920×1080 array). The micromirrors deflect at angles of $\pm 12°$, such that "on" and "off" are only $\pm 24°$. The mirrors rotate diagonally, and thus the DMD array must be mounted at a $45°$ angle in order for all the beams to stay on the plane of the table. This is shown in Fig. 10.6.

Ideally, the DMD is implemented in a holographic fashion to create an arbitrary beam shape while simultaneously cancelling aberrations [25, 35]. Accordingly, we would like to locally control both the amplitude and the phase simultaneously. This has been successfully demonstrated in the group of Markus Greiner by placing the DMD in the Fourier plane [35]. We have begun to employ a similar technique, as

Fig. 10.6 The mounting scheme used for our DMD and its attached circuitry. Even though the array of micromirrors is not square, we must rotate it by 45° such that the "on" and "off" components both stay on the plane of the table

shown in Fig. 10.5. An optimization and aberration-correction procedure is outlined in Ref. [35]. However, this technique is rather involved since the Fourier transform pattern on the DMD must have the correct amplitude and phase distribution.

Our initial goal is to control the spatial amplitude profile, and so we begin by using the DMD to generate arbitrary patterns and waveforms. The easiest way to do this is to place the DMD in the image plane, and map the DMD pattern onto the atoms with a de-magnifying telescope. In this way, the required feature size of the pattern on the DMD can be determined by the desired feature size on the atoms divided by the de-magnification factor. Arbitrary images can be generated in Matlab, and uploaded to the DMD as a bitmap (BMP) file in a look-up table (LUT). Thus, arbitrary images can easily be loaded to the DMD. We operate the DMD at the blazing angle to maximize the available power [35].

Figure 10.7 shows patterns uploaded to the DMD, and subsequently imaged onto a beam profile CCD with a magnification of 1/6.7. There are three effects in these images which are immediately apparent. First, the intensity profile of the illuminated regions is not perfectly homogeneous. This effect can be corrected as discussed in the following paragraph. Second, the images are tilted at an angle of a few degrees. This is partially due to the beam incident on the DMD, which has an angle of a few

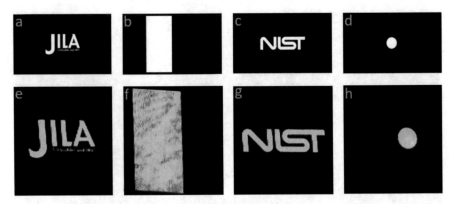

Fig. 10.7 Patterns on the DMD imaged onto a beam profile CCD. (**a**)–(**d**) Patterns uploaded to the DMD. The arrays in these images are 1920×1080 to match the size of the DMD array. (**e**)–(**h**) The DMD imaged onto a beam profile CCD using a magnification of 1/6.7. The homogeneity of the light field is set by the size of the beam illuminating the DMD. Therefore, ideally the beam is much larger than the DMD array

degrees with respect to the table. This effect causes the flat-top disk in Fig. 10.7h to be slightly elliptical. Further, since the DMD is rotated at $45°$ the profile CCD must also be oriented at this angle. The tilt angle of the images is also partially caused by a relative angle between the DMD and CCD. Third, there are high frequency fringes across the illuminated regions. These are caused by the infinitely sharp edges on the DMD which generate sinc profiles in the image plane. Essentially, this is a resolution limitation where the achromatic lens that collimates the k-vectors from the DMD has a finite NA. The lens aperture must be considered in the Fourier transform of the DMD pattern (for more discussion on Fourier optics, see Chaps. 6 and 7).

This latter problem can be addressed by using a higher power lens immediately after the DMD, but also by using a grey-scale on the DMD to provide edges of finite width. This can be done by choosing a magnification such that n^2 micromirrors map onto the diffraction-limited resolution area R^2_{min} of the objective in the plane of the atoms. Therefore, the number of micromirrors within this region which are "on" can be varied, giving a grey-scale with resolution n^2. When the value of n is chosen optimally, the edges can still be sufficiently flat for most applications while the flat-top is much more uniform.

However, for certain applications the sharpness of the edge is the most important figure of merit. Specifically, when using the light to impart an AC Stark shift for spatially resolved hyperfine (rotational) transitions for atoms (molecules), the presence of light at that lattice site is binary. The exact intensity is typically not crucial if the Stark shift is large enough to sufficiently spectrally resolve "light" and "dark" sites. For such cases, the goal is rather to have a well-defined edge between "light" and "dark" on neighboring sites. Figure 10.7b, d, f, h shows such a sharp edge as a line and a circle.

References

1. K. Aikawa, D. Akamatsu, M. Hayashi, K. Oasa, J. Kobayashi, P. Naidon, T. Kishimoto, M. Ueda, S. Inouye, Coherent transfer of photoassociated molecules into the rovibrational ground state. Phys. Rev. Lett. **105**, 203001 (2010)
2. J.-y. Choi, S. Hild, J. Zeiher, P. Schauß, A. Rubio-Abadal, T. Yefsah, V. Khemani, D.A. Huse, I. Bloch, C. Gross, Exploring the many-body localization transition in two dimensions. Science, **352**(6293), 1547–1552 (2016)
3. L. Chomaz, L. Corman, T. Yefsah, R. Desbuquois, J. Dalibard, Absorption imaging of a quasi-two-dimensional gas: a multiple scattering analysis. New J. Phys. **14**(5), 055001 (2012)
4. A. Chotia, B. Neyenhuis, S.A. Moses, B. Yan, J.P. Covey, M. Foss-Feig, A.M. Rey, D.S. Jin, J. Ye, Long-lived dipolar molecules and feshbach molecules in a 3D optical lattice. Phys. Rev. Lett. **108**, 080405 (2012)
5. J.P. Covey, L. De Marco, K. Matsuda, W. Tobias, G. Valtolina, D.S. Jin, J. Ye, A new apparatus for enhanced optical and electric manipulation of ultracold KRb molecules (2018, in preparation)
6. J.P. Covey, S.A. Moses, J. Ye, D.S. Jin, Controlling a quantum gas of polar molecules in an optical lattice, in *Royal Society of Chemistry book chapter: "Low Temperature and Low Energy Molecular Scattering"* (RSC Publishing, 2017)
7. J.P. Covey, L. De Marco, O.L. Acevedo, A.M. Rey, J. Ye, An approach to spin-resolved molecular gas microscopy. New J. Phys. **20**, 043031 (2018)
8. M.H.G. de Miranda, A. Chotia, B. Neyenhuis, D. Wang, G. Quéméner, S. Ospelkaus, J.L. Bohn, J. Ye, D.S. Jin. Controlling the quantum stereodynamics of ultracold bimolecular reactions. Nat. Phys., **7**(6), 502–507 (2011)
9. R. Desbuquois, L. Chomaz, T. Yefsah, J. Leonard, J. Beugnon, C. Weitenberg, J. Dalibard, Superfluid behaviour of a two-dimensional Bose gas. Nat. Phys. **8**, 645–648 (2012)
10. M. Feld, B. Frohlich, E. Vogt, M. Koschorreck, M. Kohl, Observation of a pairing pseudogap in a two-dimensional Ferm gas. Nature **480**, 75–78 (2011)
11. N. Gemelke, X. Zhang, C.-L. Hung, C. Chin, *In situ* observation of incompressible Mott-insulating domains in ultracold atomic gases. Nature **460**, 995–998 (2009)
12. M. Guo, B. Zhu, B. Lu, X. Ye, F. Wang, R. Vexiau, N. Bouloufa-Maafa, G. Quéméner, O. Dulieu, D. Wang, Creation of an ultracold gas of ground-state dipolar ^{23}Na^{87}Rb molecules. Phys. Rev. Lett. **116**, 205303 (2016)
13. K.R.A. Hazzard, S.R. Manmana, M. Foss-Feig, A.M. Rey, Far-from-equilibrium quantum magnetism with ultracold polar molecules. Phys. Rev. Lett. **110**, 075301 (2013)
14. K. Hueck, N. Luick, L. Sobirey, J. Siegl, T. Lompe, H. Moritz, Two-dimensional homogeneous fermi gases (2017). arXiv:1704.06315v2
15. K. Hueck, N. Luick, L. Sobirey, J. Siegl, T. Lompe, H. Moritz, L.W. Clark, C. Chin, Calibrating high intensity absorption imaging of ultracold atoms. Opt. Express **25**(8), 8670–8679 (2017)
16. C.-L. Hung, X. Zhang, L.-C. Ha, S.-K. Tung, N. Gemelke, C. Chin, Extracting densitydensity correlations from in situ images of atomic quantum gases. New J. Phys. **13**(7), 075019 (2011)
17. S. Kotochigova, D. DeMille, Electric-field-dependent dynamic polarizability and state-insensitive conditions for optical trapping of diatomic polar molecules. Phys. Rev. A **82**, 063421 (2010)
18. A. Mazurenko, C.S. Chiu, G. Ji, M. Parsons, M. Kanász-Nagy, R. Schmidt, F. Grusdt, E. Demler, D. Greif, M. Greiner, Experimental realization of a long-range antiferromagnet in the Hubbard model with ultracold atoms (2016). arXiv:1612.08436
19. D. Mitra, P.T. Brown, P. Schauß, S.S. Kondov, W.S. Bakr, Phase separation and pair condensation in a spin-imbalanced 2d fermi gas. Phys. Rev. Lett. **117**, 093601 (2016)
20. P.K. Molony, P.D. Gregory, Z. Ji, B. Lu, M.P. Köppinger, C.R. Le Sueur, C.L. Blackley, J.M. Hutson, S.L. Cornish, Creation of Ultracold ^{87}Rb^{133}Cs Molecules in the Rovibrational Ground State. Phys. Rev. Lett. **113**, 255301 (2014)

21. S.A. Moses, J.P. Covey, M.T. Miecnikowski, D.S. Jin, J. Ye, New frontiers for quantum gases of polar molecules. Nat. Phys. **13**, 13–20 (2017)

22. B. Neyenhuis, B. Yan, S.A. Moses, J.P. Covey, A. Chotia, A. Petrov, S. Kotochigova, J. Ye, D.S. Jin, Anisotropic polarizability of ultracold polar $^{40}K^{87}Rb$ molecules. Phys. Rev. Lett. **109**, 230403 (2012)

23. K.-K. Ni, S. Ospelkaus, M.H.G. de Miranda, A. Pe'er, B. Neyenhuis, J.J. Zirbel, S. Kotochigova, P.S. Julienne, D.S. Jin, J. Ye, A high phase-space-density gas of polar molecules. Science **322**(5899), 231–235 (2008)

24. K.-K. Ni, S. Ospelkaus, D. Wang, G. Quemener, B. Neyenhuis, M.H.G. de Miranda, J.L. Bohn, J. Ye, D.S. Jin. Dipolar collisions of polar molecules in the quantum regime. Nature **464**, 1324–1328 (2010)

25. A.T. Papageorge, A.J. Kollár, B.L. Lev, Coupling to modes of a near-confocal optical resonator using a digital light modulator. Opt. Express **24**(11), 11447–11457 (2016)

26. J.W. Park, S.A. Will, M.W. Zwierlein, Ultracold Dipolar Gas of Fermionic $^{23}Na^{40}K$ Molecules in their absolute ground state. Phys. Rev. Lett. **114**, 205302 (2015)

27. P.M. Preiss, R. Ma, M. Eric Tai, A. Lukin, M. Rispoli, P. Zupancic, Y. Lahini, R. Islam, M. Greiner, Strongly correlated quantum walks in optical lattices. Science **347**(6227), 1229–1233 (2015)

28. T. Takekoshi, L. Reichsöllner, A. Schindewolf, J.M. Hutson, C.R. Le Sueur, O. Dulieu, F. Ferlaino, R. Grimm, H.-C. Nägerl, Ultracold dense samples of dipolar RbCs molecules in the Rovibrational and hyperfine ground state. Phys. Rev. Lett. **113**, 205301 (2014)

29. N.V. Vitanov, A.A. Rangelov, B.W. Shore, W. Bergmann, Stimulated Raman adiabatic passage in physics, chemistry and beyond (2016). arXiv:1605.00224v2

30. D. Wang, B. Neyenhuis, M.H.G. de Miranda, K.-K. Ni, S. Ospelkaus, D.S. Jin, J. Ye, Direct absorption imaging of ultracold polar molecules. Phys. Rev. A **81**, 061404 (2010)

31. C. Weitenberg, Single-atom resolved imaging and manipulation in an atomic mott insulator, PhD thesis, Ludwig-Maximilians-Universität München, 2011

32. N.Y. Yao, C.R. Laumann, S. Gopalakrishnan, M. Knap, M. Müller, E.A. Demler, M.D. Lukin, Many-body localization in dipolar systems. Phys. Rev. Lett. **113**, 243002 (2014)

33. B. Zhu, G. Quéméner, A.M. Rey, M.J. Holland, Evaporative cooling of reactive polar molecules confined in a two-dimensional geometry. Phys. Rev. A **88**, 063405 (2013)

34. J.J. Zirbel, K.-K. Ni, S. Ospelkaus, J.P. D'Incao, C.E. Wieman, J. Ye, D.S. Jin. Collisional stability of fermionic Feshbach Molecules. Phys. Rev. Lett. **100**, 143201 (2008)

35. P. Zupancic, P.M. Preiss, R. Ma, A. Lukin, M. Eric Tai, M. Rispoli, R. Islam, M. Greiner, Ultra-precise holographic beam shaping for microscopic quantum control. Opt. Express **24**(13), 13881–13893 (2016)

Chapter 11
Outlook

With all the tools in place in the new apparatus, I now turn to the more distant future of this experiment and the next set of experiments that we have in mind. Most of these ideas have been alluded to throughout this thesis, and they include spin-orbital coupling with lattice-confined polar molecules, realizing the full spin-1/2 Hamiltonian, and demonstrating a spin-resolved quantum gas microscope. The future of ultracold polar molecules is very bright, and there are of course many proposals that I will not mention. Finally, I will discuss ideas to go beyond ITO coatings in the future in search of a more robust solution for transparent conductive coatings. Then I will conclude this thesis.

11.1 Spin-Orbital Coupling with Lattice-Confined Polar Molecules

Spin-orbital coupling has been realized in many alkali atomic systems [2, 29, 32], and typically it is induced using Raman beams to couple momentum states with internal hyperfine states. However, this generally results in large spontaneous emission and heating rates [7]. Notable exceptions include alkaline earth optical lattice clocks, where ultranarrow clock transitions can be used to eliminate heating and elucidate many-body effects [26]. Ultracold polar molecules can also be used to realize spin-orbital coupling directly through the spin–exchange interaction, and therefore without coupling to electronic excited states and thus no spontaneous emission. Furthermore, the presence of long-range dipolar interactions provides a rich, novel many-body nature to the spin-orbital coupling [48].

© Springer Nature Switzerland AG 2018
J. P. Covey, *Enhanced Optical and Electric Manipulation of a Quantum Gas of KRb Molecules*, Springer Theses, https://doi.org/10.1007/978-3-319-98107-9_11

This novel spin-orbital coupling mechanism arises from additional terms in the dipolar spin-1/2 Hamiltonian which was discussed in Chaps. 5 and 10. The splitting between the $|1, -1\rangle$, $|1, 0\rangle$, and $|1, 1\rangle$ states eliminates these terms, but the generalized dipolar interaction between molecules in an optical lattice can be written as [34, 48]

$$V_{dd} = -\frac{\sqrt{6}}{r_{ij}^3} \sum_{q=-2}^{q=2} (-1)^q C_{-q}^2(\theta, \phi) T_q^2(\mathbf{d}_i, \mathbf{d}_j), \tag{11.1}$$

where

$$C_q^k(\theta, \phi) = \sqrt{\frac{4\pi}{2k+1}} Y_{kq}(\theta, \phi) \tag{11.2}$$

where Y_{pq} is the spherical harmonic. The T_q is an irreducible tensor operator [34, 48]:

$$T_{\pm 2}^2 = \hat{d}_i^\pm \hat{d}_j^\pm, \tag{11.3}$$

$$T_{\pm 1}^2 = \frac{1}{\sqrt{2}} \left(\hat{d}_i^0 \hat{d}_j^\pm + \hat{d}_i^\pm \hat{d}_j^0 \right), \tag{11.4}$$

$$T_0^2 = \frac{1}{\sqrt{6}} \left(\hat{d}_i^+ \hat{d}_j^- + \hat{d}_i^- \hat{d}_j^+ + 2\hat{d}_i^0 \hat{d}_j^0 \right), \tag{11.5}$$

where $\hat{d}_i$ is the spin-1/2 dipole operator on site i. The final term gives rise to the spin-1/2 Hamiltonian in Chaps. 5 and 10.

To realize the spin-orbital interactions inherent in these terms, we must work in a regime where the $m_N = \pm 1$ projections of $N = 1$ are degenerate. This can be accomplished with AC or DC stark shifts that move the three uniquely, but the most straightforward approach is to use magnetic fields which Zeeman shift the nuclear degree of freedom. Accordingly, the states become resonant at 1260 G [48], and thus the $q = \pm 2$ term can be exploited. This term corresponds to a spin-exchange process which does not conserve the angular momentum of the system: $|1, -1\rangle|0, 0\rangle \rightarrow |0, 0\rangle|1, 1\rangle$, as shown in Fig. 11.1a. The Hamiltonian that governs these exchange processes has the form

$$V_{dd}^{q=\pm 2} = \frac{-3}{2r_{ij}^3} \left(\hat{d}_i^+ \hat{d}_j^+ e^{-2i\phi_{ij}} + \hat{d}_i^- \hat{d}_j^- e^{2i\phi_{ij}} \right) \sin^2 \theta_{ij}. \tag{11.6}$$

Since this process does not conserve the angular momentum of an individual pair of molecules, a net circulation in the spin current develops. The $e^{\pm 2i\phi_{ij}}$ factors give rise to a chiral excitation with d-wave symmetry, called a "chiron" (see Fig. 11.1b,

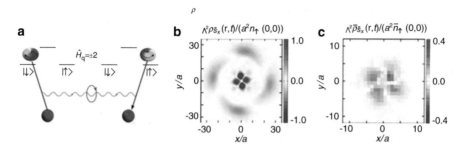

Fig. 11.1 Spin-Orbital Coupling with polar molecules in an optical lattice. (**a**) The scheme using several states for generating spi-exchange processes that do not conserve angular momentum. (**b**) The density profile of spins ranging between $|\downarrow\rangle = -1$ and $|\uparrow\rangle = 1$ for positions x and y in the 2D plane in units of lattice sites. (**c**) The same plot as in (**b**) except with filling fraction of 10%, which shows that the chiron behavior is still apparent. Reproduced from [48]

c), which has a nontrivial Berry phase of 2π [48]. Because of the $\sin^2\theta_{ij}$ factor it is most straightforward to perform this experiment in a single plane perpendicular to the quantization axis such that $\theta = \pi/2$. The initialization of this experiment is very similar to the spin-impurity dynamics experiments discussed in Chap. 10. After an arbitrary spin impurity region has been created in the $|1, -1\rangle$ state, the magnetic field can quickly be ramped to 1260 G, which is within reach of the bias coils in the new apparatus. Then, the spin-exchange process can proceed by both $|1, -1\rangle|0, 0\rangle \rightarrow |0, 0\rangle|1, -1\rangle$ and $|1, -1\rangle|0, 0\rangle \rightarrow |0, 0\rangle|1, 1\rangle$. The latter gives rise to the development of chirons.

The observation of such phenomena should be relatively straightforward since the chiron is a topological excitation and thus robust to system imperfections, and it is even expected to persist to low filling fractions [48]. In order to prevent spin-exchange processes from occurring before the excited spin projections are degenerate, we can leave the spin addressing beam on after the localized spin excitation. The AC stark shift from the addressing beam is very large compared to $J_\perp$, and thus any dynamics will be off-resonantly suppressed until the beam is turned off. We anticipate that our tools of high resolution in situ detection and nearly-single-spin addressing in an optical lattice combined with a bias field of 1260 G will allow us to observe spin-orbital coupling, which will be especially clear given our current filling fraction of 30% and our anticipated filling fraction of 50% after a few improvements.

These tools could be extended to the observation of emergent Weyl fermionic excitations [49]. Weyl fermions are massless chiral particles, such as the chirons that arise from the dipolar spin-orbital couplings. To observe Weyl excitations we need a weakly broken time-reversal symmetry, such as a large magnetic field (already present). This opens intriguing prospects of observing chiral anomaly, non-local electrodynamics, and non-Anderson disorder-driven transitions [49]. The Weyl character of the excitations could be probed with a Ramsey protocol, but requires pulses of interfering Raman beams [49]. Such a procedure should give rise to the observation of Weyl nodes through in situ detection [49].

11.2 Quantum Magnetism in the XXZ Hamiltonian

The ability to apply large, stable electric fields opens the door to possible experiments with both the Ising term and the spin-exchange term, i.e. the XXZ Hamiltonian, whose relative strengths depend on the field as shown in Fig. 7.6. The ability to tune $J_\perp$ and J_z allows us to access an enormous wealth of intriguing phenomena. For example, when $J_\perp = J_z$, dynamics are governed by the Heisenberg Hamiltonian, which can be used to study critical points and phase transitions of magnetic systems. It is expected that there is a phase transition between an XY paramagnetic spin fluid when $J_\perp$ dominates (small electric field) and a Ne'el-ordered antiferromagnet when J_z dominates (large electric field) [23]. Thus, at sufficiently large fields we may expect spin-exchange oscillations to diminish and give way to the proliferation of spin correlations, which could be observed with high resolution in situ imaging. However, the phase diagram depends on dimensionality and is in general poorly understood beyond 1D where the system is highly frustrated. Our ability to tune the dimensionality of the system offers even more information, particularly since the dipolar interactions are anisotropic.

Moreover, it is possible to study critical phenomena by quenching the system across the continuous quantum phase transition. However, this requires the system to be near the ground state, and to have nearly 100% filling. By jumping the electric field from $J_\perp \gg J_z$ to $J_z \gg J_\perp$ at a rate large compared to $J_\perp$, J_z, we can access the Kibble-Zurek mechanism which arises from spontaneously breaking a global symmetry [25, 56]. Here we would expect to see the formation of topological defects such as domain structures, whose size scales with the ramp rate across the transition in accordance with the Kibble-Zurek scaling [24]. The density scale (inverse length scale) of such domains obeys a power law in the quench rate of the electric field across the transition, and the power exponent is given in terms of the critical exponents of the transition.

11.3 An Approach to Molecular Gas Microscopy in an Optical Lattice

Efforts towards in situ single molecule detection in both optical lattices [16, 19] and optical tweezers [22, 31] are increasingly active. It is thus timely to consider single-site microscopy of polar molecules, and how two rotational states can be simultaneously detected. The latter point is of particular importance for molecules due to the limited fidelity of their creation process, which results in the occupation of a given site being a priori unknown [11, 35]. The ability to detect the presence or absence of a molecule on a given site, in addition to its state, is therefore important.

In this section, I present a technique by which two molecular rotational states can be unambiguously detected, giving simultaneously site-resolved and spin-resolved detection. After discussing the approach in which molecules are detected, I describe how these techniques can be extended to fluorescence detection for molecular gas microscopy. Finally, I present simulations of out-of-equilibrium spin systems, and describe how the investigation of such dynamics would proceed under a spin-resolved molecular gas microscope. ^{40}K^{87}Rb [36] is used as an example throughout, but these methods are general, and they can be applied to other ultracold bialkali molecular species [20, 33, 39, 46, 50].

11.3.1 Spin-Resolved Imaging Protocol

Formation and dissociation of ground-state molecules in an optical lattice is well understood [10, 11, 35], and it has been demonstrated that either K or Rb atoms from dissociated KRb molecules can be imaged in situ to yield consistent results [11, 35]. This is an important step towards building a spin-resolved microscope of polar molecules as this requires both atoms to be a suitable proxy for molecules. It has also been shown, equally as important, that one species can be removed with resonant light without deleterious effects on the other. In fact, removing a unique atomic species is straightforward since the difference in transition frequencies is on the order of several Thu.

Accordingly, each spin state can be mapped onto either atomic species, making a spin-resolved microscope of KRb molecules equivalent to simultaneous microscopes of K [8, 15, 21] and Rb [3, 47]. We also note that KRb has the convenient feature that the two $S_{1/2} \to P_{3/2}$ atomic transitions are separated by only 13 nm, which minimizes chromatic aberrations in imaging while still making wavelength separation straightforward.

Before describing the mapping protocol in detail, we note that there is freedom in which molecular state is mapped onto which atomic species. Either option is possible in general, but for a given molecular species one choice may prove to be advantageous over the other. In what follows, we focus on the case in which $| \downarrow \rangle$ is mapped onto K, as shown in Fig. 11.2. We also note that we assume the lattice is sufficiently deep such that the tunneling probabilities for molecules or atoms are negligible throughout the entirety of this procedure. We also note that we have only one molecule per site during the experiment and mapping protocol [55].

Step I: Dissociating $| \downarrow \rangle$ Molecules

The first step in the mapping protocol is to convert $| \downarrow \rangle$ molecules to a pair of K and Rb on the same site. This is done by reversing the STIRAP sequence, which

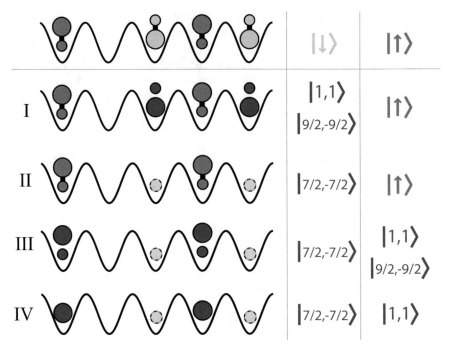

Fig. 11.2 Various steps required to realize a spin-resolved microscope, mapping $|\downarrow\rangle$ onto K and $|\uparrow\rangle$ onto Rb. Reproduced from [12]

uniquely couples the $|\downarrow\rangle$ state to Feshbach molecules, and then sweeping the magnetic field across the Feshbach resonance to dissociate the Feshbach molecules. The fidelity of this process is $\sim$90–95% [11, 36], which is limited by the STIRAP efficiency. After dissociation, Rb is in the $|F, m_F\rangle = |1, 1\rangle$ state and K is in the $|9/2, -9/2\rangle$ state, as shown in step I of Fig. 11.2 (F denotes the total angular momentum, and m_F is its projection on the magnetic field axis).

This STIRAP pulse will not couple $|\uparrow\rangle$ molecules to the specific Feshbach state, but we must ensure that it does not couple them to any other state. Accordingly, we must look carefully at the full quantum state of the molecules, including the nuclear hyperfine degrees of freedom [37]. In a STIRAP sequence with identically polarized beams, the molecule's angular momentum is conserved, which eliminate most pathways. Furthermore, the STIRAP linewidth is $\sim$200 kHz and the sequence is 5 μs long [36], and since the rotational splitting is on the order of GHz, the spectral resolution prohibits driving to an uncoupled state. An electric field can additionally be used during STIRAP to provide even greater selectivity [36].

Step II: Removing Rb Atoms

Following dissociation, one of the species must be removed such that each doublon becomes a site with a single atom. As discussed above, this can be done with a pulse of resonant light. Such pulses, which have been demonstrated to preserve the other atomic spin with high fidelity, are routinely used in atomic quantum gas microscope experiments, with typical blast times between $10\,\mu s$ and $1\,ms$ depending on the species and lattice depth [3, 4, 8, 9, 15, 21, 35, 40, 47]. Moreover, pulses resonant with the atomic transition have a negligible effect on molecules [11, 35, 36], thus leaving the $|\uparrow\rangle$ molecules unaffected.

The choice of which atom to be removed first depends on the particular molecular species. Here I consider the removal of Rb first, as in Fig. 11.2. Effective removal of Rb can be accomplished in $\sim 1\,ms$ with $>99.9\%$ fidelity by simultaneously driving the $|1, 1\rangle \rightarrow |2, 2\rangle'$ and $|2, 2\rangle \rightarrow |3, 3\rangle'$ transitions, where $|F, m_F\rangle'$ denotes the electronically excited $P_{3/2}$ state. This leaves behind K atoms in the $|9/2, -9/2\rangle$ state.

With Rb atoms removed, we must flip the K atoms to a different spin state such that they can be distinguished from the K atoms that emerge in the subsequent steps. For K, a convenient transition is from $|9/2, -9/2\rangle$ to $|7/2, -7/2\rangle$, which has a transition frequency of $\sim 2.7\,GHz$ under 550 G; this transition can be driven with $>99\%$ fidelity in tens of μs [9]. As shown in Fig. 11.2, after step II, molecules in the $|\uparrow\rangle$ state and unpaired K atoms remain, with the latter depicted in lighter color with hashed edges to signify the $|7/2, -7/2\rangle$ state.

Step III: Dissociating $|\uparrow\rangle$ Molecules

Next we convert the $|\uparrow\rangle$ molecules back to a doublon, first by applying a rotational-state-changing microwave pulse to transfer $|\uparrow\rangle \rightarrow |\downarrow\rangle$. Then the same reverse-STIRAP and magneto-dissociation sequence converts the $|\uparrow\rangle$ molecules to doublons. As before, the doublons are in the states $|1, 1\rangle$ for Rb and $|9/2, -9/2\rangle$ for K. This is shown in step III of Fig. 11.2.

Step IV: Removing K Atoms

In the final step, K atoms from the doublons are removed, which shows why the atomic state transfer was necessary in step II. The K atoms that remain from the previous step are in the $|7/2, -7/2\rangle$ state, which are sufficiently detuned from the $|9/2, -9/2\rangle \rightarrow |11/2, -11/2\rangle'$ transition as to not be affected. Therefore, the K atoms in the doublon can be removed with a blast pulse in $<1\,ms$ with high fidelity [9], which completes the mapping procedure.

11.3.2 Alternative Protocol

The particular choice of mapping protocol hinges on the fact that two atoms on a lattice site will quickly undergo inelastic loss when one occupies an excited hyperfine level [11]. Therefore, the time spent in this configuration must be minimized. Due to the inverted hyperfine structure of K, the ground hyperfine level is the cycling state, and thus it is convenient to remove K atoms in Step IV as opposed to Rb. In cases where neither atom has an inverted hyperfine structure, such as NaRb or RbCs, the protocol with higher overall fidelity would involve removing the lighter atom in Step IV.

11.3.3 Detection Fidelity and Technical Requirements

After the above protocol, the sites that were $| \downarrow \rangle$ molecules are K atoms in $|7/2, -7/2\rangle$ and the sites that were $| \uparrow \rangle$ molecules are Rb atoms in $|1, 1\rangle$. The subsequent imaging via optical molasses [3, 47] cooling or Raman sideband cooling [41] (RSC) operates primarily on the transitions $|2, 2\rangle \rightarrow |3, 3\rangle'$ for Rb and $|9/2, -9/2\rangle \rightarrow |11/2, -11/2\rangle'$ for K, but the repump transitions $|1, 1\rangle \rightarrow |2, 2\rangle'$ for Rb and $|7/2, -7/2\rangle \rightarrow |9/2, -9/2\rangle'$ for K are required anyway to keep atoms in the cycling transition [3, 8, 15, 21, 47]. Hence, both transitions are driven in parallel, and the initial states of the K and Rb atoms in the lattice are unimportant for imaging.

It is critical that the desired atomic species is preserved during removal of the other. One possible deleterious effect is heating of the desired species by the removal pulses. The heating rates are calculated using the photon scattering rates given by:

$$\Gamma_{sc} = \left(\frac{\Gamma}{2}\right) \frac{(I/I_{sat})}{1 + 4(\Delta/\Gamma)^2 + (I/I_{sat})}, \tag{11.7}$$

where Γ is the linewidth of the electronic excited state, I is the intensity of the removal beam, I_{sat} is the saturation intensity of the atomic transition, and Δ is the detuning from resonance of the excited state. The heating rate can be calculated from this scattering rate via $\dot{T} = 1/3 \times (2 \times E_{rec})/k_B \times \Gamma_{sc}$, where the $1/3$ is specific to the case of a harmonic trap, and $2 \times E_{rec}$ is the recoil energy of photon absorption and re-emission [18].

For a removal beam of intensity $I = I_{sat}$ and detuning $\Delta = 0$ for the target species, the scattering rate is $\Gamma_{sc} \approx 10^7$ s^{-1}, which corresponds to a resonant heating rate of $\approx 1 \mu$K/μs for Rb in the $|1, 1\rangle$ state and $\approx 2 \mu$K/μs for K in $|9/2, -9/2\rangle$. These on-resonance heating rates are strong such that the removal pulse can be quite short, on the scale of 1 ms or less, and consequently do not perturb the desired species or spin states for which the off-resonance heating rates are very low. During Step II, for example, a resonant pulse to remove Rb atoms will have no impact on either K atoms or KRb molecules. In Step IV, we need to remove K atoms in the $|9/2, -9/2\rangle$ state without affecting the $|7/2, -7/2\rangle$ state. For the removal light tuned on resonance with $|9/2, -9/2\rangle \to |11/2, -11/2\rangle'$, the $|7/2, -7/2\rangle \to |9/2, -9/2\rangle'$ transition has the smallest detuning of 0.84 GHz, corresponding to a scattering rate of $\Gamma_{|7/2,-7/2\rangle \to |9/2,-9/2\rangle'} = 245$ s^{-1}. Meanwhile, a pulse of length ~ 1 ms is more than sufficient to remove $|9/2, -9/2\rangle$ K atoms.

Another factor that may play a role during the removal pulses is photo-association. Quantum gas microscopes operate with parity detection because every pair of atoms located on a single lattice site is associated into an electronically excited molecule during the roughly second-long interrogation time. However, the probability of this process happening during the $\sim$ms removal pulses is known to be low.

I can now estimate the overall detection fidelity of each individual spin component. The detection of $|\downarrow\rangle$ relies on STIRAP, Rb removal, and a K spin flip, which have respective efficiencies of 95%, >99%, and >99%. This corresponds to an overall fidelity of 93%. The detection of $|\uparrow\rangle$ relies on STIRAP and K removal (while preserving K atoms of the other hyperfine state), which have respective efficiencies of 95% and >95% [9]. However, it also relies on minimal deleterious effects from the first STIRAP pulse. Such effects are expected to be at the single percent level, but will vary between molecular species. Therefore we estimate the overall fidelity of $|\uparrow\rangle$ detection to be 85–90%.

11.4 Spin Impurity Dynamics

It is not always reliable to trust intuition regarding the dynamics of a long-range quantum many-body spin system. The interplay between the range of interactions and the dimensionality of a system can have enormous qualitative effects on dynamics. In this section, I numerically investigate the spin model that governs systems of polar molecules, and present simulations of spin-exchange dynamics mediated by long-range dipolar interactions. These simulations further motivate the need for specific spin patterns and for spin-resolved microscopy.

The Hamiltonian that describes this exchange process is [17, 51, 53]:

$$\hat{H} = \frac{\hbar J}{2} \sum_{j=1}^{\mathcal{N}} \sum_{l<j} (1 - 3\cos^2\theta_{lj}) ||\boldsymbol{r}_l - \boldsymbol{r}_j||^{-3} \left(\hat{s}_l^+ \hat{s}_j^- + \hat{s}_l^- \hat{s}_j^+ \right), \tag{11.8}$$

where $\hat{s}_j$ are spin–1/2 operators, $\boldsymbol{r}_j$ is the position vector of molecule i in the lattice, $\mathcal{N}$ is the number of molecules, and θ_{lj} is the angle between the vector $\boldsymbol{r}_l - \boldsymbol{r}_j$ and the quantization axis, which is defined by an external magnetic field. This Hamiltonian is derived for the case of zero electric field at DC so that there is no Ising term proportional to $\hat{s}_l^z \hat{s}_j^z$. The exchange coupling J is determined by the transition dipole moment, $d_{\uparrow\downarrow}$, between two selected rotational states and is given by $\hbar J = d_{\uparrow\downarrow}^2/(4\pi\epsilon_0 a^3)$, where ϵ_0 is the permittivity of free space and a is the lattice spacing.

For KRb molecules in a three-dimensional lattice with spacing $a = 532\,\text{nm}$, using the two rotational states $|0, 0\rangle$ and $|1, 0\rangle$, $d_{\uparrow\downarrow} \approx -0.57/\sqrt{3}$ D, and $|J|/2 = 2\pi \times 104\,\text{Hz}$. For $|0, 0\rangle$ and $|1, -1\rangle$, $|J|/2 = 2\pi \times 52\,\text{Hz}$. Assume that the quantization axis is perpendicular to the two-dimensional plane of molecules, which implies that $\cos\theta_{lk} = 0$, resulting in an isotropic interaction. Disorder in such a model manifests itself in two ways: either a disordered potential landscape can be added to the lattice using a projected potential as discussed in the previous section; or, since the lattice filling fraction is less than unity, there is a natural disorder in the filling arrangement of the molecules and consequently in the $||\boldsymbol{r}_l - \boldsymbol{r}_j||^{-3}$ geometrical prefactors for dipolar coupling. The question about the existence of many-body localization in this 2D system remains open and largely unexplored [5, 6, 27, 28, 42, 54], while the single-particle (Anderson localization) case is better understood [14, 52].

The initial state is prepared by driving all molecules to the $|\downarrow\rangle$ state and then selecting an excitation region of the lattice in which the molecules are excited to $|\uparrow\rangle$. Once prepared, the total spin $\hat{S}_z$ remains constant since it is a conserved quantity. In this case, a useful observable for quantifying the spin dynamics is the *imbalance* $\hat{I}_z(t)$, which is a measure of the average magnetization of the gas weighted by the initial on-site values, defined as

$$\hat{I}_z(t) = \frac{4}{\mathcal{N}} \sum_{j=1}^{\mathcal{N}} \langle \hat{s}_j^z(0) \rangle \hat{s}_j^z(t). \tag{11.9}$$

The imbalance is equal to one initially and will either decay if the system is ergodic or will remain finite at long times if the system localizes.

Since the exact simulation of interacting spins in a 2D lattice becomes intractable as the number of molecules is increased, the simulations presented here are based on the discrete truncated Wigner approximation (DTWA) [43–45]. The DTWA is a semi-classical method that models the dynamics through a set of classical trajectories (in this case Bloch vector dynamics) that evolve according to the mean-field equations. The random initial conditions for each of these trajectories are selected according to the initial state represented as a quasi-probability Wigner distribution. Despite its semiclassical character, DTWA has been shown to be capable of reproducing quantum correlations and to capture the quantum spin dynamics beyond the mean-field limit; it has been demonstrated recently to be useful for exploring ergodic/localized dynamics [1].

I now consider the case where the initial excitation is a checkerboard pattern (which can be generated with spatial light modulators as discussed in Chap. 10), where the size of each square can be as small as 2×2 sites. For large checker sizes the spin imbalance stays near unity even out to long times, in accordance with the

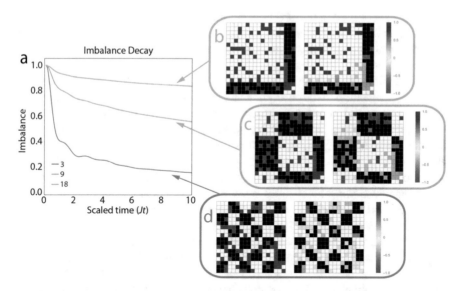

Fig. 11.3 Numerical DTWA simulations of spin dynamics. (**a**) The spin imbalance as a function of time for 36×36 sites for checker sizes of 3×3 (blue), 9×9 (orange), and 18×18 (green) sites. (**b**)–(**d**) Spin-resolved microscopy of an 18×18 subset with 25% filling at $t = 0$ (left) and $Jt = 10$ (right) corresponding to the three curves in (**a**). Sites on checkers that are initially $| \uparrow\rangle$ ($| \downarrow\rangle$) are colored white (black). The color scale is from blue to red for $| \uparrow\rangle$ to $| \downarrow\rangle$, respectively. Reproduced from [12]

case of a single large circular $| \uparrow\rangle$ region. However, as the checker size decreases, substantial spin dynamics occurs and the imbalance drops to ≈ 0.2 within $t \approx 2/J$, as in Fig. 11.3a.

Snapshots of the spin distributions at $t = 0$ and $t = 10/J$ for checker sizes of 3×3, 9×9, and 18×18 are shown for Fig. 11.3b–d. It is also worth noting that under our initial conditions total spin observables $\hat{S}_\alpha$ ($\alpha = x, y, z$) have expectation values equal to zero throughout the entire dynamic process. Therefore, site- and spin-resolved detection is essential to glean all the information from such dynamics. These numerical predictions provide a strong motivation for the experimental capability of microscopic, spin-resolved detection since substantial diffusion occurs only for microscopic areas.

11.5 Beyond ITO

In the last section of this thesis I will discuss possible alternatives to ITO coatings, which were shown in Chap. 8 to have a number of limitations for high voltage applications as well as alkali adsorption. While the longevity of this coating is not known at the time of this writing, I would except that it will cause some of the problems experienced in the next several years. Therefore it is worth pursuing other options, although the problems with large electric fields and the destructive nature of high voltage conditioning will likely be general problems that plague any thin film technology.

As I mentioned, other options include metal coatings such as nickel or copper, but these have far too much absorption and reflection. Another option could be thin films of carbon nanotubes [38], which are being explored in Stuttgart for thermal vapor cells of Rydberg atoms (T. Pfau, private communication). Similarly, graphene thin films are an intriguing option. It is possible that such coats will have lower absorption and reflection. Further it is possible, even likely, that they will be more stable in the presence of alkali vapor. However, the surface figure may be worse in carbon nanotube thin film coatings since they are essentially a spaghetti of carbon nanotubes, and so surface features may exceed $\lambda/10 \approx 80$ nm at which point they may affect very high resolution imaging systems.

It is unclear whether these coatings will have better performance in high voltage applications than ITO. The coating may be more chemically robust, but if it is less smooth it will be more susceptible to electron emission. Further, the work function for carbon nanotube thin films is slightly lower than for ITO [30]. In any case, many tests would need to be performed in a similar fashion to the tests we did with ITO.

Other intriguing options include a much thicker ITO coating than we are using, such as 100 nm, and then small regions of thickness ≈ 10 nm where the beams pass through the coating. A nickel coating could also be used in such a way, or graphene thin films. As was discussed in Chap. 10, the lattice beam is at a $\approx 10°$ angle, and so the regions of thinner coating would be offset from the cloud by a few mm. Higher absorption and reflection will not severely affect the high resolution imaging system, and it will simply mean that the integration time must be a few percent longer to achieve the same number of photons collected on the camera.

11.6   Conclusions

Before 2012 the future of ultracold polar molecules was shrouded by their complexity and slow progress. The molecular gases were too hot, the motion in the trap was not entirely controlled, their interaction with the trapping light was poorly understood, and the molecules suffered fast loss from chemical reactions. Further, large electric fields only enhanced the chemical loss in 3D gases. This was only beginning to be controlled using 2D systems [13]. Moreover, there was a quickly growing number of theoretical proposals for ultracold polar molecule experiments, but many of them required the tools which were developed during the time of my thesis. My PhD work has contributed to bringing us from thermal $(T > T_F)$ samples of ultracold polar KRb molecules in 2D or 3D all the way to a spin-exchange-interacting, low-entropy gas of molecules in an optical lattice where advanced AC and DC electric field tools and high resolution imaging and addressing tools are beginning to be used to create a molecular gas microscope.

More generally, the field of ultracold polar molecules has changed enormously in the past 6 years. Since 2014, there are now many ultracold molecule experiments in the world. These other experiments are moving very quickly, but they all have more to learn about their molecular species. We felt that it was time to implement a new generation of our apparatus which was sufficiently complex to go beyond the capabilities of any other experiment. Our design was intended to include every tool we wanted, and in the end we were able to successfully combine all of these tools together into the same apparatus. We hope that our new apparatus will serve as a standard for what is possible with ultracold polar molecules as we continue to pursue novel directions in the coming years.

References

1. O.L. Acevedo, A. Safavi-Naini, J. Schachenmayer, M.L. Wall, R. Nandkishore, A.M. Rey, Exploring many-body localization and thermalization using semiclassical methods. Phys. Rev. A **96**(3), 033604 (2017)
2. M. Aidelsburger, M. Atala, M. Lohse, J.T. Barreiro, B. Paredes, I. Bloch, Realization of the Hofstadter hamiltonian with ultracold atoms in optical lattices. Phys. Rev. Lett. **111**, 185301 (2013)
3. W.S. Bakr, J.I. Gillen, A. Peng, S. Folling, M. Greiner, A quantum gas microscope for detecting single atoms in a Hubbard-regime optical lattice. Nature **462**, 74–77 (2009)
4. P.T. Brown, D. Mitra, E. Guardado-Sanchez, P. Schauß, S.S. Kondov, E. Khatami, T. Paiva, N. Trivedi, D.A. Huse, W. Bakr, Observation of canted antiferromagnetism with ultracold fermions in an optical lattice. arXiv:1612.07746 (2016)
5. A.L. Burin, Localization in a random XY model with long-range interactions: Intermediate case between single-particle and many-body problems. Phys. Rev. B **92**(10), 104428 (2015)
6. A.L. Burin, Many-body delocalization in a strongly disordered system with long-range interactions: finite-size scaling. Phys. Rev. B **91**(9), 094202 (2015)
7. L.W. Cheuk, A.T. Sommer, Z. Hadzibabic, T. Yefsah, W.S. Bakr, M.W. Zwierlein, Spin-injection spectroscopy of a spin-orbit coupled fermi gas. Phys. Rev. Lett. **109**, 095302 (2012)
8. L.W. Cheuk, M.A. Nichols, M. Okan, T. Gersdorf, V.V. Ramasesh, W.S. Bakr, T. Lompe, M.W. Zwierlein, Quantum-gas microscope for fermionic atoms Phys. Rev. Lett. **114**, 193001 (2015)
9. L.W. Cheuk, M.A. Nichols, K.R. Lawrence, M. Okan, H. Zhang, E. Khatami, N. Trivedi, T. Paiva, M. Rigol, M.W. Zwierlein, Observation of spatial charge and spin correlations in the 2D Fermi-Hubbard model. Science **353**(6305), 1260–1264 (2016)
10. A. Chotia, B. Neyenhuis, S.A. Moses, B. Yan, J.P. Covey, M. Foss-Feig, A.M. Rey, D.S. Jin, J. Ye, Long-lived dipolar molecules and Feshbach molecules in a 3D optical lattice. Phys. Rev. Lett. **108**, 080405 (2012)
11. J.P. Covey, S.A. Moses, M. Garttner, A. Safavi-Naini, M.T. Miecnkowski, Z. Fu, J. Schachenmayer, P.S. Julienne, A.M. Rey, D.S. Jin, J. Ye, Doublon dynamics and polar molecule production in an optical lattice. Nat. Commun. **7**, 11279 (2016)
12. J.P. Covey, L. De Marco, O.L. Acevedo, A.M. Rey, J. Ye, An approach to spin-resolved molecular gas microscopy. New J. Phys. **20**, 043031 (2018)
13. M.H.G. de Miranda, A. Chotia, B. Neyenhuis, D. Wang, G. Quéméner, S. Ospelkaus, J.L. Bohn, J. Ye, D.S. Jin, Controlling the quantum stereodynamics of ultracold bimolecular reactions. Nat. Phys. **7**(6), 502–507 (2011)
14. X. Deng, B.L. Altshuler, G.V. Shlyapnikov, L. Santos, Quantum levy flights and multifractality of dipolar excitations in a random system. Phys. Rev. Lett. **117**, 020401 (2016)

15. G.J.A. Edge, R. Anderson, D. Jervis, D.C. McKay, R. Day, S. Trotzky, J.H. Thywissen, Imaging and addressing of individual fermionic atoms in an optical lattice. Phys. Rev. A **92**, 063406 (2015)

16. M.W. Gempel, T. Hartmann, T.A. Schulze, K.K. Voges, A. Zenesini, S. Ospelkaus, Versatile electric fields for the manipulation of ultracold NaK molecules. New J. Phys. **18**(4), 045017 (2016)

17. A.V. Gorshkov, S.R. Manmana, G. Chen, J. Ye, E. Demler, M.D. Lukin, A.M. Rey, Tunable superfluidity and quantum magnetism with ultracold polar molecules. Phys. Rev. Lett. **107**, 115301 (2011)

18. R. Grimm, M. Weidemller, Y.B. Ovchinnikov, Optical dipole traps for neutral atoms, in *Advances in Atomic, Molecular, and Optical Physics*, vol. 42 (Springer, Berlin, 2000), pp. 95–170

19. M. Gröbner, P. Weinmann, F. Meinert, K. Lauber, E. Kirilov, H.-C. Nägerl, A new quantum gas apparatus for ultracold mixtures of K and Cs and KCs ground-state molecules. J. Mod. Opt. 1–11 (2016)

20. M. Guo, B. Zhu, B. Lu, X. Ye, F. Wang, R. Vexiau, N. Bouloufa-Maafa, G. Quéméner, O. Dulieu, D. Wang, Creation of an ultracold gas of ground-state dipolar ^{23}Na^{87}Rb molecules. Phys. Rev. Lett. **116**, 205303 (2016)

21. E. Haller, J. Hudson, A. Kelly, D.A. Cotta, B. Peaudecerf, G.D. Bruce, S. Kuhr. Single-atom imaging of fermions in a quantum-gas microscope. Nat. Phys. **11**, 738–742 (2015)

22. N.R. Hutzler, L.R. Liu, Y. Yu, K.-K. Ni, Eliminating light shifts in single-atom optical traps. arXiv:1605.09422v1 (2016)

23. R. Jafari, A. Langari, Phase diagram of the one-dimensional $s = \frac{1}{2}$ xxz model with ferromagnetic nearest-neighbor and antiferromagnetic next-nearest-neighbor interactions. Phys. Rev. B **76**, 014412 (2007)

24. D. Jaschke, K. Maeda, J. D. Whalen, M. L. Wall, and L. D. Carr. Critical phenomena and Kibble-Zurek scaling in the long-range quantum Ising chain. New J. Phys. **19**, 033032 (2017)

25. T.W.B. Kibble, Topology of cosmic domains and strings. J. Phys. A Math. Gen. **9**(8), 1387 (1976)

26. S. Kolkowitz, S.L. Bromley, T. Bothwell, M.L. Wall, G.E. Marti, A.P. Koller, X. Zhang, A.M. Rey, J. Ye, Spin-orbit-coupled fermions in an optical lattice clock. Nature **542**, 66–70 (2017)

27. M.P. Kwasigroch, N.R. Cooper, Bose-Einstein condensation and many-body localization of rotational excitations of polar molecules following a microwave pulse. Phys. Rev. A **90**, 021605 (2014)

28. M.P. Kwasigroch, N.R. Cooper, Synchronization transition in dipole-coupled two-level systems with positional disorder, Phys. Rev. A **96**, 053610 (2017)

29. Y. Lin, R.L. Compton, K. Jimenez-Gercia, J.V. Porto, I.B. Spielman, Synthetic magnetic fields for ultracold neutral atoms. Nature **462**, 628–632 (2009)

30. P. Liu, Q. Sun, F. Zhu, K. Liu, K. Jiang, L. Liu, Q. Li, S. Fan, Measuring the work function of carbon nanotubes with thermionic method. Nano Lett. **8**(2), 647–651 (2008)

31. L.R. Liu, J.T. Zhang, Y. Yu, N.R. Hutzler, T. Rosenband, K.-K. Ni, Ultracold molecular assembly. arXiv:1605.09422v1 (2017)

32. H. Miyake, G.A. Siviloglou, C.J. Kennedy, W.C. Burton, W. Ketterle, Realizing the harper hamiltonian with laser-assisted tunneling in optical lattices. Phys. Rev. Lett. **111**, 185302 (2013)

33. P.K. Molony, P.D. Gregory, Z. Ji, B. Lu, M.P. Köppinger, C.R. Le Sueur, C.L. Blackley, J.M. Hutson, S.L. Cornish, Creation of ultracold ^{87}Rb^{133}Cs molecules in the Rovibrational ground state. Phys. Rev. Lett. **113**, 255301 (2014)

34. S.A. Moses, A quantum gas of polar molecules in an optical lattice. PhD thesis, University of Colorado, Boulder (2016)

35. S.A. Moses, J.P. Covey, M.T. Miecnikowski, B. Yan, B. Gadway, J. Ye, D.S. Jin, Creation of a low-entropy quantum gas of polar molecules in an optical lattice. Science **350**(6261), 659–662 (2015)

36. K.-K. Ni, S. Ospelkaus, M.H.G. de Miranda, A. Pe'er, B. Neyenhuis, J.J. Zirbel, S. Kotochigova, P.S. Julienne, D.S. Jin, J. Ye, A high phase-space-density gas of polar molecules. Science **322**(5899), 231–235 (2008)
37. S. Ospelkaus, K.-K. Ni, G. Quéméner, B. Neyenhuis, D. Wang, M.H.G. de Miranda, J.L. Bohn, J. Ye, D.S. Jin, Controlling the hyperfine state of Rovibronic ground-state polar molecules. Phys. Rev. Lett. **104**, 030402 (2010)
38. A.K. Pal, R.K. Roy, S.K. Mandal, S. Gupta, B. Deb, Electrodeposited carbon nanotube thin films. Thin Solid Films **476**(2), 288–294 (2005)
39. J.W. Park, S.A. Will, M.W. Zwierlein, Ultracold dipolar gas of fermionic $^{23}Na^{40}K$ molecules in their absolute ground state. Phys. Rev. Lett. **114**, 205302 (2015)
40. M.F. Parsons, A. Mazurenko, C.S. Chiu, G. Ji, D. Greif, M. Greiner, Site-resolved measurement of the spin-correlation function in the Fermi-Hubbard model. Science **353**(6305), 1253–1256 (2016)
41. Y.S. Patil, S. Chakram, L.M. Aycock, M. Vengalattore, Nondestructive imaging of an ultracold lattice gas. Phys. Rev. A **90**, 033422 (2014)
42. D. Peter, S. Müller, S. Wessel, H.P. Büchler, Anomalous behavior of spin systems with dipolar interactions. Phys. Rev. Lett. **109**, 025303 (2012)
43. L. Pucci, A. Roy, M. Kastner, Simulation of quantum spin dynamics by phase space sampling of Bogoliubov-Born-Green-Kirkwood-Yvon trajectories. Phys. Rev. B **93**(17), 174302 (2016)
44. J. Schachenmayer, A. Pikovski, A.M. Rey, Dynamics of correlations in two-dimensional quantum spin models with long-range interactions: a phase-space Monte-Carlo study. New J. Phys. **17**(6), 065009 (2015)
45. J. Schachenmayer, A. Pikovski, A.M. Rey, Many-body quantum spin dynamics with Monte carlo trajectories on a discrete phase space. Phys. Rev. X **5**(1), 011022 (2015)
46. F. Seeßelberg, N. Buchheim, Z.-K. Lu, T. Schneider, X.-Y. Luo, E. Tiemann, I. Bloch, C. Gohle, Modeling the adiabatic creation of ultracold, polar $^{23}Na^{40}K$ molecules. arXiv:1709.00902v1 (2017)
47. J.F. Sherson, C. Weitenberg, M. Endres, M. Cheneau, I. Bloch, S. Kuhr, Single-atom-resolved fluorescence imaging of an atomic Mott insulator. Nature **467**, 68–72 (2010)
48. S.V. Syzranov, M.L. Wall, V. Gurarie, A.M. Rey, Spin–orbital dynamics in a system of polar molecules. Nat. Commun. **5**, 5391 (2014)
49. S.V. Syzranov, M.L. Wall, B. Zhu, V. Gurarie, A.M. Rey, Emergent Weyl excitations in systems of polar particles. Nat. Commun. **7**, 13543 (2016)
50. T. Takekoshi, L. Reichsöllner, A. Schindewolf, J.M. Hutson, C.R. Le Sueur, O. Dulieu, F. Ferlaino, R. Grimm, H.-C. Nägerl. Ultracold dense samples of dipolar RbCs molecules in the rovibrational and hyperfine ground state. Phys. Rev. Lett. **113**, 205301 (2014)
51. M.L. Wall, K.R.A. Hazzard, A.M. Rey, Quantum magnetism with ultracold molecules, In *From Atomic to Mesoscale*, Chap. 1 (World Scientific, Singapore, 2015), pp. 3–37
52. T. Xu, R.V. Krems, Quantum walk and Anderson localization of rotational excitations in disordered ensembles of polar molecules. New J. Phys. **17**(6), 065014 (2015)
53. B. Yan, S.A. Moses, B. Gadway, J.P. Covey, K.R.A. Hazzard, A.M. Rey, D.S. Jin, J. Ye, Observation of dipolar spin-exchange interactions with lattice-confined polar molecules. Nature **501**(7468), 521–525 (2013)
54. N.Y. Yao, C.R. Laumann, S. Gopalakrishnan, M. Knap, M. Müller, E.A. Demler, M.D. Lukin, Many-body localization in dipolar systems. Phys. Rev. Lett. **113**, 243002 (2014)
55. B. Zhu, B. Gadway, M. Foss-Feig, J. Schachenmayer, M.L. Wall, K.R.A. Hazzard, B. Yan, S.A. Moses, J.P. Covey, D.S. Jin, J. Ye, M. Holland, A.M. Rey, Suppressing the loss of ultracold molecules via the continuous quantum Zeno effect. Phys. Rev. Lett. **112**, 070404 (2014)
56. W.H. Zurek, Cosmological experiments in superfluid helium? Nature **317**, 505–508 (1985)

Bibliography

1. W.S. Bakr, A. Peng, M.E. Tai, R. Ma, J. Simon, J.I. Gillen, S. Fölling, L. Pollet, M. Greiner, Probing the superfluid–to–Mott insulator transition at the single-atom level. Science **329**(5991), 547–550 (2010)
2. M.A. Baranov, H. Fehrmann, M. Lewenstein, Wigner crystallization in rapidly rotating 2D dipolar fermi gases. Phys. Rev. Lett. **100**, 200402 (2008)
3. J. Baron, W.C. Campbell, D. DeMille, J.M. Doyle, G. Gabrielse, Y.V. Gurevich, P.W. Hess, N.R. Hutzler, E. Kirilov, I. Kozyryev, B.R. O'Leary, C.D. Panda, M.F. Parsons, E.S. Petrik, B. Spaun, A.C. Vutha, A.D. West, Order of magnitude smaller limit on the electric dipole moment of the electron. Science **343**(6168), 269–272 (2014)
4. J.F. Barry, D.J. McCarron, E.B. Norrgard, M.H. Steinecker, D. Demille, Magneto-optical trapping of a diatomic molecule. Nature **512**, 286–289 (2014)
5. M. Boll, T.A. Hilker, G. Salomon, A. Omran, J. Nespolo, L. Pollet, I. Bloch, C. Gross, Spin- and density-resolved microscopy of antiferromagnetic correlations in fermi-Hubbard chains. Science **353**(6305), 1257–1260 (2016)
6. M. Cheneau, P. Barmettler, D. Poletti, M. Endres, P. Schauss, T. Fukuhara, C. Gross, I. Bloch, C. Kollath, S. Kuhr, Light-cone-like spreading of correlations in a quantum many-body system. Nature **481**, 484–487 (2012)
7. L.W. Cheuk, M.A. Nichols, K.R. Lawrence, M. Okan, H. Zhang, M.W. Zwierlein, Observation of 2D fermionic mott insulators of ^{40}K with single-site resolution. Phys. Rev. Lett. **116**, 235301 (2016)
8. S. Choi, J. Choi, R. Landig, G. Kucsko, H. Zhou, J. Isoya, F. Jelezko, S. Onoda, H. Sumiya, V. Khemani, C. von Keyserlingk, N.Y. Yao, E. Demler, M.D. Lukin, Observation of discrete time-crystalline order in a disordered dipolar many-body system. arXiv:1609.08684 (2016)
9. T.D. Cumby, R.A. Shewmon, M.-G. Hu, J.D. Perreault, D.S. Jin, Feshbach-molecule formation in a Bose-Fermi mixture. Phys. Rev. A **87**, 012703 (2013)
10. W. Dowd, R.J. Roy, R.K. Shrestha, A. Petrov, C. Makrides, S. Kotochigova, S. Gupta, Magnetic field dependent interactions in an ultracold Li-Yb($^3P_2$) mixture. New J. Phys. **17**(5), 055007 (2015)
11. M. Endres, M. Cheneau, T. Fukuhara, C. Weitenberg, P. Schauß, C. Gross, L. Mazza, M.C. Bañuls, L. Pollet, I. Bloch, S. Kuhr, Observation of correlated particle-hole pairs and string order in low-dimensional Mott insulators. Science **334**(6053), 200–203 (2011)
12. M. Endres, H. Bernien, A. Keesling, H. Levine, E.R. Anschuetz, A. Krajenbrink, C. Senko, V. Vuletic, M. Greiner, M.D. Lukin, Atom-by-atom assembly of defect-free one-dimensional cold atom arrays. Science (2016)

© Springer Nature Switzerland AG 2018

J. P. Covey, *Enhanced Optical and Electric Manipulation of a Quantum Gas of KRb Molecules*, Springer Theses, https://doi.org/10.1007/978-3-319-98107-9

13. J.P. Gaebler, J.T. Stewart, J.L. Bohn, D.S. Jin, p-wave Feshbach molecules. Phys. Rev. Lett. **98**, 200403 (2007)

14. J. Goldwin, S. Inouye, M.L. Olsen, B. Newman, B.D. DePaola, D.S. Jin, Measurement of the interaction strength in a Bose-Fermi mixture with ^{87}Rb and ^{40}K. Phys. Rev. A **70**, 021601 (2004)

15. D. Greif, M.F. Parsons, A. Mazurenko, C.S. Chiu, S. Blatt, F. Huber, G. Ji, M. Greiner, Site-resolved imaging of a fermionic Mott insulator. Science **351**(6276), 953–957 (2016)

16. M. Greiner, S. Folling, Condensed-matter physics: optical lattices. Nature **453**, 736–738 (2008)

17. T. Grünzweig, A. Hilliard, M. McGovern, M.F. Andersen, Near-deterministic preparation of a single atom in an optical microtrap. Nat. Phys. **6**, 951–954 (2010)

18. K.R.A. Hazzard, A.V. Gorshkov, A.M. Rey, Spectroscopy of dipolar fermions in layered two-dimensional and three-dimensional lattices. Phys. Rev. A **84**, 033608 (2011)

19. T.A. Hilker, G. Salomon, F. Grusdt, A. Omran, M. Boll, E. Demler, I. Bloch, C. Gross, Revealing hidden antiferromagnetic correlations in doped Hubbard chains via string correlators. arXiv:1702.00642v1 (2017)

20. J.J. Hudson, D.M. Kara, I.J. Smallman, B.E. Sauer, M.R. Tarbutt, E.A. Hinds, Improved measurement of the shape of the electron. Nature **460**, 995–998 (2009)

21. M.T. Hummon, M. Yeo, B.K. Stuhl, A.L. Collopy, Y. Xia, J. Ye, 2D magneto-optical trapping of diatomic molecules. Phys. Rev. Lett. **110**, 143001 (2013)

22. S. Inouye, J. Goldwin, M.L. Olsen, C. Ticknor, J.L. Bohn, D.S. Jin, Observation of heteronuclear Feshbach resonances in a mixture of bosons and fermions. Phys. Rev. Lett. **93**, 183201 (2004)

23. L. Isenhower, E. Urban, X.L. Zhang, A.T. Gill, T. Henage, T.A. Johnson, T.G. Walker, M. Saffman, Demonstration of a neutral atom controlled-not quantum gate. Phys. Rev. Lett. **104**, 010503 (2010)

24. R. Islam, R. Ma, P.M. Preiss, M.E. Tai, A. Lukin, M. Rispoli, M. Greiner, Measuring entanglement entropy in a quantum many-body system. Nature **528**, 77–83 (2015)

25. P.S. Julienne, T.M. Hanna, Z. Idziaszek, Universal ultracold collision rates for polar molecules of two alkali-metal atoms. Phys. Chem. Chem. Phys. **13**, 19114–19124 (2011)

26. A.M. Kaufman, B.J. Lester, C.A. Regal, Cooling a single atom in an optical tweezer to its quantum ground state. Phys. Rev. X **2**, 041014 (2012)

27. A.M. Kaufman, B.J. Lester, C.M. Reynolds, M.L. Wall, M. Foss-Feig, K.R.A. Hazzard, A.M. Rey, C.A. Regal, Two-particle quantum interference in tunnel-coupled optical tweezers. Science **345**(6194), 306–309 (2014)

28. S.S. Kondov, W.R. McGehee, J.J. Zirbel, B. DeMarco, Three-dimensional Anderson localization of ultracold matter. Science **334**(6052), 66–68 (2011)

29. M. Kwon, M.F. Ebert, T.G. Walker, M. Saffman, Parallel low-loss measurement of multiple atomic qubits. arXiv:1706.09497v1 (2017)

30. B.J. Lester, A.M. Kaufman, C.A. Regal, Raman cooling imaging: detecting single atoms near their ground state of motion. Phys. Rev. A **90**, 011804 (2014)

31. B.J. Lester, N. Luick, A.M. Kaufman, C.M. Reynolds, C.A. Regal, Rapid production of uniformly filled arrays of neutral atoms. Phys. Rev. Lett. **115**, 073003 (2015)

32. R.M. Lutchyn, E. Rossi, S.D. Sarma, Spontaneous interlayer superfluidity in bilayer systems of cold polar molecules. Phys. Rev. A **82**, 061604 (2010)

33. K.D. Nelson, X. Li, D.S. Weiss, Imaging single atoms in a three-dimensional array. Nat. Phys. **3**, 556–560 (2007)

34. D.J. Nesbitt, Toward state-to-state dynamics in ultracold collisions: lessons from high-resolution spectroscopy of weakly bound molecular complexes. Chem. Rev. **112**(9), 5062–5072 (2012)

35. F. Nogrette, H. Labuhn, S. Ravets, D. Barredo, L. Béguin, A. Vernier, T. Lahaye, A. Browaeys, Single-atom trapping in holographic 2d arrays of microtraps with arbitrary geometries. Phys. Rev. X **4**, 021034 (2014)

36. A. Omran, M. Boll, T.A. Hilker, K. Kleinlein, G. Salomon, I. Bloch, C. Gross, Microscopic observation of Pauli blocking in degenerate Fermionic lattice gases. Phys. Rev. Lett. **115**, 263001 (2015)

37. M.F. Parsons, F. Huber, A. Mazurenko, C.S. Chiu, W. Setiawan, K. Wooley-Brown, S. Blatt, M. Greiner, Site-resolved imaging of Fermionic ^{6}Li in an optical lattice. Phys. Rev. Lett. **114**, 213002 (2015)

38. B. Pasquiou, A. Bayerle, S.M. Tzanova, S. Stellmer, J. Szczepkowski, M. Parigger, R. Grimm, F. Schreck, Quantum degenerate mixtures of strontium and rubidium atoms. Phys. Rev. A **88**, 023601 (2013)

39. P.M. Preiss, R. Ma, M.E. Tai, J. Simon, M. Greiner, Quantum gas microscopy with spin, atom-number, and multilayer readout. Phys. Rev. A **91**, 041602 (2015)

40. C.A. Regal, C. Ticknor, J.L. Bohn, D.S. Jin, Tuning p-wave interactions in an ultracold fermi gas of atoms. Phys. Rev. Lett. **90**, 053201 (2003)

41. T. Rom, T. Best, O. Mandel, A. Widera, M. Greiner, T.W. Hänsch, I. Bloch, State selective production of molecules in optical lattices. Phys. Rev. Lett. **93**, 073002 (2004)

42. P. Schauß, J. Zeiher, T. Fukuhara, S. Hild, M. Cheneau, T. Macrì, T. Pohl, I. Bloch, C. Gross, Crystallization in Ising quantum magnets. Science **347**(6229), 1455–1458 (2015)

43. M. Schlosser, J. Kruse, C. Gierl, S. Teichmann, S. Tichelmann, G. Birkl, Fast transport, atom sample splitting and single-atom qubit supply in two-dimensional arrays of optical microtraps. New J. Phys. **14**(12), 123034 (2012)

44. E.S. Shuman, J.F. Barry, D.R. Glenn, D. DeMille, Radiative force from optical cycling on a diatomic molecule. Phys. Rev. Lett. **103**, 223001 (2009)

45. J. Simon, W.S. Bakr, R. Ma, M.E. Tai, P.M. Preiss, M. Greiner, Quantum simulation of antiferromagnetic spin chains in an optical lattice. Nature **472**, 307–312 (2011)

46. E.W. Streed, J. Mun, M. Boyd, G.K. Campbell, P. Medley, W. Ketterle, D.E. Pritchard, Continuous and pulsed quantum Zeno effect. Phys. Rev. Lett. **97**, 260402 (2006)

47. B.K. Stuhl, M.T. Hummon, M. Yeo, G. Quemener, J.L. Bohn, J. Ye, Evaporative cooling of the dipolar hydroxyl radical. Nature **492**, 396–400 (2012)

48. B.K. Stuhl, B.C. Sawyer, D. Wang, J. Ye, Magneto-optical trap for polar molecules. Phys. Rev. Lett. **101**, 243002 (2008)

49. M.R. Tarbutt, T.C. Steimle, Modeling magneto-optical trapping of CaF molecules. Phys. Rev. A **92**, 053401 (2015)

50. N.Y. Yao, A.C. Potter, I.-D. Potirniche, A. Vishwanath, Discrete time crystals: rigidity, criticality, and realizations. Phys. Rev. Lett. **118**, 030401 (2017)

51. M. Yeo, M.T. Hummon, A.L. Collopy, B. Yan, B. Hemmerling, E. Chae, J.M. Doyle, J. Ye, Rotational state microwave mixing for laser cooling of complex diatomic molecules. Phys. Rev. Lett. **114**, 223003 (2015)

52. J. Zhang, P.W. Hess, A. Kyprianidis, P. Becker, A. Lee, J. Smith, G. Pagano, I.-D. Potirniche, A.C. Potter, A. Vishwananth, N.Y. Yao, C. Monroe, Observation of a discrete time crystal. arXiv:1609.08684 (2016)

53. V. Zhelyazkova, A. Cournol, T.E. Wall, A. Matsushima, J.J. Hudson, E.A. Hinds, M.R. Tarbutt, B.E. Sauer, Laser cooling and slowing of CaF molecules. Phys. Rev. A **89**, 053416 (2014)

Printed in the United States
By Bookmasters